C0-BZC-085

OXFORD MEDICAL PUBLICATIONS

Psychopharmacology and Food

BRITISH ASSOCIATION
FOR PSYCHOPHARMACOLOGY MONOGRAPHS

1. Psychopharmacology of affective disorders
edited by E. S. Paykel and A. Coppen

2. Psychopharmacology of anticonvulsants
edited by Merton Sandler

3. Psychopharmacology of old age
edited by David Wheatley

4. Psychopharmacology and sexual disorders
edited by David Wheatley

5. Psychopharmacology of the limbic system
edited by Michael R. Trimble and E. Zarifian

6. Psychopharmacology: recent advances and future prospects
edited by Susan D. Iversen

7. Psychopharmacology and food
edited by Merton Sandler and Trevor Silverstone

Psychopharmacology and Food

BRITISH ASSOCIATION
FOR PSYCHOPHARMACOLOGY
MONOGRAPH
No. 7

EDITED BY

MERTON SANDLER

AND

TREVOR SILVERSTONE

Oxford New York Tokyo
OXFORD UNIVERSITY PRESS
1985

Oxford University Press, Walton Street, Oxford OX2 6DP
Oxford New York Toronto
Delhi Bombay Calcutta Madras Karachi
Kuala Lumpur Singapore Hong Kong Tokyo
Nairobi Dar es Salaam Cape Town
Melbourne Auckland
and associated companies in
Beirut Berlin Ibadan Nicosia

Oxford is a trade mark of Oxford University Press

British Library Cataloguing in Publication Data

Psychopharmacology and food. — (A British Association for Psychopharmacology monograph; no. 7) — (Oxford medical publications)
1. Nutrition 2. Psychopharmacology
I. Sandler, Merton II. Silverstone, Trevor
III. Series
613.2 RA784

ISBN 0-19-261458-4

Library of Congress Cataloging in Publication Data

Psychopharmacology and food. — (British Association for Psychopharmacology monograph; no. 7) — (Oxford medical publications)
Collection of papers, based on a symposium held in London, in December 1983, under the auspices of the British Association for Psychopharmacology.
Includes bibliographies and index.
1. Psychopharmacology—Congresses. 2. Appetite—Effect of drugs on—Congresses. 3. Nutritionally induced diseases—Congresses.
I. Sandler, Merton. II. Silverstone, Trevor. III. British Association for Psychopharmacology. IV. Series. V. Series: Oxford medical publications.

[DNLM: 1. Appetite—drug effects—congresses. 2. Appetite Depressants—Congresses. 3. Appetite Disorders—drug therapy—congresses. 4. Feeding Behavior—drug effects—congresses. 5. Food—adverse effects—congresses. 6. Obesity—drug therapy—congresses. 7. Psychopharmacology—congresses.
W1 BR343D v.7 / W1 102 P974 1983]
RM315.P749784 1985 615'.78 85-7096
ISBN 0-19-261458-4

Typeset by Rod-Art, Abingdon, Oxon.
Printed in Great Britain by J. W. Arrowsmith Ltd, Bristol

'reface

:he good Lord', as Brillat-Savarin reminds us, 'makes us eat in order to live, and ı doing so tempts us through appetite and rewards us through pleasure'. This ook, which we have entitled, *Psychopharmacology and food*, does two things: first of all examines the ways in which drugs can influence the appetite, and ıereby affect physiological processes underlying the intake of food; second, it ɔnsiders the effects that food itself can have on the body. Thus it covers both ıe *pharmacology of feeding* and *the action of foodstuffs as pharmacological ;ents* – hence the title.

Appetite, or 'the desire for food', being a subjective phenomenon influenced y, but not synonymous with, hunger, is largely controlled by processes occur-ıg in the central nervous system. So we begin with a set of three reviews ɔvering the underlying neurophysiology and neurochemistry of eating. Edmund .olls gives an account of his recent work unravelling the complex and fas-inating interaction in the brains of primates of food-related cues, nutritional atus, and neuronal activity. Steven Cooper goes on to describe how neuro-eptides act centrally and peripherally and therby influence feeding and drinking ı laboratory animals and in human subjects. The third chapter in this opening :ction is a consideration by Gerald Curzon of the mechanisms by which food ıd food deprivation may affect transmitter synthesis.

The following five chapters relate to the effects of a variety of drugs on ormal feeding and the application of these drugs in clinical practice. John lundell sets the scene with a comprehensive review of the ways in which :ntrally acting drugs can influence food intake by affecting a number of neuro-ansmitter systems in the brain, which interact one with the other. Jozsef Knoll ıen describes his efforts to isolate a naturally occurring anorexigenic substance hich, if successful, could revolutionize our treatment of appetite disorders. ntil then, we are confined to those synthetic products currently available. John lunro, taking a hard look at the use of appetite-suppressant drugs in the eatment of obesity, concludes that they can be of definite value when used :nsibly in patients for whom there are real clinical indications. As a practising sychiatrist, Trevor Silverstone was struck by how frequently patients receiving ɔng-term treatment with antipsychotic drugs become extremely fat. In his ıapter he discusses the many ways in which drugs, used primarily for their :fect on the mental state of the patients for whom they are prescribed, can also gnificantly affect their appetite and body weight. The final chapter in this :ction is by Gerald Russell who discusses the place of drugs in the treatment of ıorexia nervosa and of that seemingly new scourge of modern womanhood, ulimia nervosa.

The last section contains three chapters relating to some of the effects foo themselves can have on bodily processes in health and disease. David Shaw loo at the nutritional basis for dementia. Merton Sandler summarizes his extensi researches into nutritional triggers of migraine. Maurice Lessof draws the whc proceedings to a close with a succinct survey of the various ways in which foo can act to bring about those bodily changes already described.

We have put together this collection of papers, based on a highly successf symposium held in London in December 1983 under the auspices of the Briti Association for Psychopharmacology, in the hope that it will be of value to for whom food and feeding has more than a merely hedonistic interest. T areas covered include many which are only sparsely considered elsewhere. Th we believe that this book will be of real value to a wide range of professiona these include physicians, neuroscientists, psychiatrists, psychologists, dietitiar and pharmacologists.

We wish to thank our fellow contributors for making the preparation of tl book such a pleasant and relatively pain-free experience.

London M.
July 1984 T.

Contents

Contributors

C. ABEL, *Medical Unit, Eastern General Hospital, Seafield Street, Edinburgh, EH6 7LN, UK.*

JOHN E. BLUNDELL, *Department of Psychology, University of Leeds, Leeds, LS2 9JT, UK.*

STEVEN J. COOPER, *Department of Psychology, University of Birmingham, PO Box 363, Birmingham, B15 2TT, UK.*

K. O. CHUNG-A-ON, *Biochemical Psychiatry Laboratory, Welsh National School of Medicine, Whitchurch Hospital, Cardiff, CF4 7XB, UK.*

G. CURZON, *Department of Neurochemistry, Institute of Neurology, Queen's Square, London, WC1, UK.*

J. W. T. DICKERSON, *Department of Biochemistry, Division of Nutrition and Food Science, University of Surrey, Guildford, GU2 5XH, UK.*

VIVETTE GLOVER, *Department of Chemical Pathology, Queen Charlotte's Maternity Hospital, Goldhawk Road, London, W6 0XG, UK.*

J. KNOLL, *Semmelweiss Orvostudomanyi Egyetem, 1085 Budapest VIII, Ulloi Ut 26, Hungary.*

M. H. LESSOF, *Department of Medicine, Guy's Hospital Medical School, London, SE1 9RT, UK.*

JULIA T. LITTLEWOOD, *Department of Chemical Pathology, Queen Charlotte's Maternity Hospital, Goldhawk Road, London, W6 0XG, UK.*

J. F. MUNRO, *Medical Unit, Eastern General Hospital, Seafield Street, Edinburgh, EH6 7LN, UK.*

EDMUND T. ROLLS, *Department of Experimental Psychology, University of Oxford, South Parks Road, Oxford, OX1 3UD, UK.*

G. F. M. RUSSELL, *Institute of Psychiatry, De Crespigny Park, Denmark Hill, London, SE5 8AI, UK.*

M. SANDLER, *Department of Chemical Pathology, Queen Charlotte's Maternity Hospital, Goldhawk Road, London W6 0XG, UK.*

D. M. SHAW, *Biochemical Psychiatry Laboratory, Department of Psychological Medicine, Welsh National School of Medicine, Whitchurch Hospital, Cardiff, CF4 7XB, UK.*

TREVOR SILVERSTONE, *Academic Unit of Human Psychopharmacology, German Hospital, Ritson Road, London, E8 1DS, UK.*

E. A. SWEENEY, *Biochemical Psychiatry Laboratory, Department of Psychological Medicine, Welsh National School of Medicine, Whitchurch Hospital, Cardiff, CF4 7XB, UK.*

D. E. THOMAS, *Biochemical Psychiatry Laboratory, Department of Psychological Medicine, Welsh National School of Medicine, Whitchurch Hospital, Cardiff, CF4 7XB, UK.*

S. F. TIDMARSH, *Biochemical Psychiatry Laboratory, Department of Psychological Medicine, Welsh National School of Medicine, Whitchurch Hospital, Cardiff, CF4 7XB, UK.*

1

The neurophysiology of feeding

EDMUND T. ROLLS

INTRODUCTION

From clinical evidence it has been known since early this century that damage to the base of the brain can influence food intake and body weight. Later it was demonstrated that one critical region is the ventromedial hypothalamus, where, in animals, bilateral lesions led to hyperphagia and obesity (see Grossman 1967, 1973; Rolls 1981*b*). Anand and Brobeck, in 1951, discovered that bilateral lesions of the lateral hypothalamus could produce a reduction in feeding and body weight. Evidence of this type led in the 1950s and 1960s to the view that food intake was controlled by two interacting 'centres', a feeding centre in the lateral hypothalamus, and a satiety centre in the ventromedial hypothalamus (Stellar 1954; Grossman 1967, 1973).

Soon, there appeared problems with this evidence for a dual centre hypothesis of the control of food intake. For example, lesions of the lateral hypothalamus which were effective in producing aphagia also damaged fibre pathways coursing nearby, such as the dopaminergic nigro-striatal bundle, and damage to these pathways outside the lateral hypothalamus could produce aphagia (Marshall, Richardson, and Teitelbaum 1974). Thus by the middle 1970s it was clear that the lesion evidence for a lateral hypothalamic feeding centre was not straightforward, because at least part of the effect of the lesions was due to damage to fibres of passage travelling through or near the lateral hypothalamus (Stricker and Zigmond 1976). However, it was not clear by this time what, if any, role the hypothalamus played in feeding. To obtain more direct evidence on the neural processing involved in feeding, the activity of single neurones in the hypothalamus and other brain regions such as the amygdala, prefrontal cortex, and striatum, is being analysed, as will be described next.

NEURONAL ACTIVITY IN THE LATERAL HYPOTHALAMUS DURING FEEDING

It has been found that there is a population of neurones in the lateral hypothalamus and substantia innominata of the monkey with responses that are related to feeding (Rolls 1981*a,b,* 1983). These neurones, which comprised 13.6 per cent in one sample of 764 hypothalamic neurones, respond to the taste and/or sight of food. The neurones respond to taste in that they respond only when certain substances, such as glucose solution but not water or saline, are in the mouth, and in that their firing rates are related to the concentration of the substance to which they respond (Rolls, Burton, and Mora 1980). These neurones did not respond simply in relation to mouth movements, and comprised 4.3 per cent of the sample of 764 neurones. The responses of the neurones associated with the sight of food occurred as soon as the monkey saw the food, before the food was in his mouth, and occurred only to foods and not to non-food objects. These neurones comprised 11.8 per cent of the sample of 764 neurones (Rolls, Burton, and Mora 1976, 1980). Some of these neurones (2.5 per cent of the total sample) responded to both the sight and taste of food (Rolls *et al.* 1976, 1980). The finding that there are neurones in the lateral hypothalamus of the monkey which respond to the sight of food has been confirmed by Ono, Nishino, Sasaki, Fukuda, and Muramoto (1980).

Effect of hunger

The responses of these neurones only occur to the sight, or to the taste, of food if the monkey is hungry (Burton, Rolls, and Mora 1976). Thus neuronal responses which occur to food in the hypothalamus depend on the motivational state of the animal. In the monkey the signals which reflect the motivational state and perform this modulation include the following. Gastric distension is one signal, as shown by the finding that after a monkey has fed to satiety, relief of gastric distension by drainage of ingested food through a gastric cannula leads to the almost immediate resumption of feeding (Gibbs, Maddison, and Rolls 1981). (Because feeding is reinstated so rapidly, it is probably due to relief of distension rather than to the altered availability of metabolites in later parts of the gut). Another signal is provided by the presence of food in the duodenum and later parts of the gut, as shown by the finding that if ingested food is allowed to drain from a duodenal cannula (situated near the pylorus), then normal satiety is not shown, and the monkey feeds almost continuously (Gibbs *et al.* 1981). Under these conditions, food does not accumulate normally in the stomach, showing that duodenal influences, or influences of more distal origin, are required to control gastric emptying, and thus to allow gastric distension to play a role in satiety. In this way an 'enterogastric loop' contributes to satiety. The presence of food in the duodenum also contributes to satiety, as shown by the finding that duodenal infusions of food at rates similar to those of gastric emptying reduce the rate of feeding (Gibbs *et al.* 1981). Other signals that influence hunger and satiety presumably reflect the metabolic state of the animal, and may include such signals as glucose and insulin levels (LeMagnen 1980; Friedman and Stricker 1976; Woods *et al.* 1980; McKay, Stein, West, and Porte 1980).

The hypothalamus is not necessarily the first stage at which processing of such environmental stimuli is modulated by hunger. To investigate whether hunger modulates neuronal responses in parts of the visual system through which visual information is likely to reach the hypothalamus (see below), the activity of neurones in the visual inferior temporal cortex was recorded in the same testing situations. Here, it was found that the neuronal responses to visual stimuli were not dependent on hunger (Rolls, Judge, and Sanghera 1977). Nor were neuronal responses in the amygdala, which connects the inferior temporal visual cortex to the hypothalamus (see below), found to depend on hunger (Sanghera, Rolls, and Roper-Hall 1979) (although there still could be other pre-hypothalamic areas in which motivational modulation of processing occurs). Thus for visual processing, neuronal responsiveness has been found to be modulated by hunger in the hypothalamus, but not at earlier stages in the processing. This is further evidence that the hypothalamic neurones which respond to food are likely to be involved in the responsiveness of the organism to food (see also Rolls 1982).

Sensory-specific modulation of the responsiveness of lateral hypothalamic neurones and of appetite

During these experiments on satiety it was observed that if a lateral hypothalamic neurone had ceased to respond to a food on which the monkey had been fed to satiety, then the neurone might still respond to a different food. This occurred for neurones with responses associated with the taste (Rolls 1981*b*) or sight (Rolls and Rolls 1982) of food. Corresponding to this neuronal specificity of the effects of feeding to satiety, the monkey rejected the food on which he had been fed to satiety, but accepted other foods which he had not previously been fed.

As a result of these neurophysiological and behavioural observations showing the specificity of satiety in the monkey, experiments were performed to determine whether satiety was specific to foods eaten in man. It was found that the pleasantness of the taste of food eaten to satiety decreased more than for foods which had not been eaten (Rolls, Rolls, Rowe, and Sweeney 1981*b*). One implication of this finding is that if one food is eaten to satiety, appetite reduction for other foods is often incomplete, and this should mean that in man also at least some of the other foods will be eaten. This has been confirmed in an experiment in which either sausages or cheese with crackers were eaten for lunch. The liking for the food eaten decreased more than for the food not eaten and, when an unexpected second course was offered, more was eaten if a subject had not been given that food in the first course than if he had been given that food in the first course (98 per cent v. 40 per cent eaten in the second courses, $p < 0.01$ (Rolls *et al.* 1981*b*). A further implication of these findings is that if a variety of foods is available, the total amount consumed will be more than when only one food is offered repeatedly. This prediction has been confirmed in a study in which humans ate more when offered a variety of sandwich fillings or a variety of types of yoghurt which differed in taste, texture and colour than when offered one filling or a single type of yoghurt (Rolls, Rowe, Rolls, Kingston, Megson, and Gunary 1981*a*). This prediction is further confirmed in a study in which humans were offered a relatively nomal meal of four courses, and it was found that the change of food at each course significantly enhanced intake (Rolls, Van Duijenvoorde, and Rolls 1984). Because sensory factors such as similarity of colour, shape, flavour, and texture are usually more important than metabolic equivalence in terms of protein, carbohydrate, and fat content in influencing how foods interact in this type of satiety, the term 'sensory-specific satiety' has come to be used (Rolls and Rolls 1977, 1982; Rolls, Rowe, and Rolls 1982; Rolls *et al.* 1981*a,b*).

The parallel between these studies of feeding in humans, and of the neurophysiology of hypothalamic neurones in the monkey, has been extended by the observations that in humans, sensory-specific satiety occurs for the sight as well as for the taste of food (Rolls, Rolls, and Rowe 1983*a*). Further, to complement the finding that in the hypothalamus neurones are found which respond differently to food and to water (see below), and that satiety with water can decrease the responsiveness of hypothalamic neurones which respond to water, it has been shown that in man motivation-specific satiety can also be detected. For example, satiety with water decreases the pleasantness of the sight and taste of water, but not of food (Rolls, Rolls, and Rowe 1983*a*).

The enhanced eating when a variety of foods is available, as a result of the operation of sensory-specific satiety, may have been advantageous in evolution in ensuring that different foods with important different nutrients were consumed, but today in man, when a wide variety of foods is readily available, it may be a factor which can lead to overeating and obesity. In a test of this hypothesis in the rat, it has been found that variety itself can lead to obesity in the rat (Rolls, Van Duijenvoorde, and Rowe 1983*b*).

Effects of learning

The responses of these hypothalamic neurones in the primate become associated as a result of learning with the sight of food. This is shown by experiments in which the neurones come to respond to the sight of a previously neutral stimulus, such as a syringe, from which the monkey is fed orally; in which the neurones cease to respond to a stimulus if it is no longer associated with food (in extinction or passive avoidance); and in which the responses of these neurones remain associated with whichever visual stimulus is associated with food reward in a visual discrimination and its reversals (Mora, Rolls, and Burton 1976). This type of learning is important because it allows organisms to respond appropriately to environmental stimuli which previous experience has shown are foods. The brain mechanisms for this type of learning are discussed below.

The responses of these neurones suggest that they are involved in responses to food. Further evidence for this is that the responses of these neurones occur with relatively short latencies of 150-200 ms, and thus precede and predict the responses of the hungry monkey to food (Rolls, Sanghera, and Roper-Hall 1979).

Evidence that the responses of hypothalamic neurones are related to the reward value of food

Given that these hypothalamic neurones respond to food when it is rewarding, that is when the animal will work to obtain food, there is a possibility that their responses are related to the reward value that food has for the hungry animal.

Evidence consistent with this comes from studies using electrical stimulation of the brain. It has been found that electrical stimulation of some brain regions is rewarding in that animals, including man, will work to obtain electrical stimulation of some sites in the brain (Olds 1977; Rolls 1975, 1976, 1979). At some sites, including the lateral hypothalamus, the electrical stimulation appears to produce reward which is equivalent to food for the hungry animal, in that the animal will work hard to obtain the stimulation if he is hungry, but will work much less for it if he has been satiated (Olds 1977; Hoebel 1969). There is even evidence that the reward at some sites can mimic food for a hungry animal, and at other sites water for a thirsty animal, in that rats chose electrical stimulation at one hypothalamic site when hungry and at a different site when thirsty (Gallistel and Beagley 1971). It was therefore interesting to discover that some of the neurones, normally activated by food when the monkey was hungry, were also activated by brain-stimulation reward (Rolls 1975, 1976; Rolls *et al*. 1980). Thus, there was convergence of the effects of natural food reward and brain-stimulation reward at some brain sites (e.g. the orbitofrontal cortex and amygdala), onto single hypothalamic neurones. Further, it was shown that self-stimulation occurred through the recording electrode if it was near a region where hypothalamic neurones that responded to food had been recorded, and that this self-stimulation was attenuated by feeding the monkey to satiety (Rolls *et al*. 1980).

The finding that these neurones were activated by brain-stimulation reward is consistent with the hypothesis that their activity is related to reward produced by food, and not to some other effect of food. Indeed, this evidence from the convergence of brain-stimulation reward and food reward onto these hypothalmic neurones, and from the self-stimulation found through the recording electrode, suggests that animals work to obtain activation of these neurones by food, and that this activation is what makes food rewarding. At the same time this accounts for self-stimulation of some brain sites, which would be understood as the animal seeking to activate the neurones that he normally seeks to activate by food when he is hungry. This, and other evidence (Rolls 1975, 1982), indicates that feeding normally occurs in order to obtain the sensory input produced by food which is rewarding if the animal is hungry.

Sites in the hypothalamus and basal forebrain of neurones that respond to food

These neurones are found as a relatively small proportion of cells in a region which includes the lateral hypothalamus and substantia innominata and extends from the lateral hypothalamus posteriorly through the anterior hypothalamus and lateral preoptic area to a region ventral to and anterior to the anterior commissure (Fig. 7 of Rolls *et al*. 1979). Useful further information about the particular populations of neurones in these regions with feeding-related activity, and about the functions of these neurones in feeding could be provided by evidence on their output connections. It is known that some hypothalamic neurones project to brainstem autonomic regions such as the dorsal motor nucleus of the vagus (Saper, Loewy, Swanson, and Cowan 1976; Saper, Swanson, and Cowan 1979). If some of the hypothalamic neurones with feeding-related activity projected in this way, it would be very likely that their functions would include the generation of autonomic responses to the sight of food. Some hypothalamic neurones project to the substantia nigra (Nauta and Domesick 1978), and some neurones in the lateral hypothalamus and basal magnocellular forebrain nuclei of Meynert project directly to the cerebral cortex (Kievit and Kuypers 1975; Divac 1975). If some of these were feeding-related neurones, then by such routes they could influence whether or not feeding was initiated. In order to determine to which regions hypothalamic neurones with feeding-related activity project, electrical stimulation is being applied to these different regions to determine from which regions hypothalamic neurones with feeding-related activity can be antidromically activated. It has so far been found in such experiments by E. T. Rolls, E. Murzi and C. Griffiths that some of these feeding-related neurones in the lateral hypothalamus and substantia innominata project directly to the cerebral cortex, to such areas as the prefrontal cortex in the sulcus principalis and the supplementary motor cortex. This provides evidence that at least some of these neurones with feeding-related activity project this information to the cerebral cortex, where it could be used in such processes as the initiation of feeding behaviour. These experiments also indicate that at least some of these feeding-related neurones are in the basal magnocellular forebrain nuclei of Meynert, which is quite consistent with the reconstructions of the recording sites (Rolls, Sanghera, and Roper-Hall 1979, Fig. 7). In addition, it seems quite likely that at least some of the feeding-related neurones influence the brainstem autonomic centres, and this remains to be investigated further.

Functions of the hypothalamus in feeding

The functions of the hypothalamus in feeding are thus related at least in part to the inputs which it receives from the forebrain, in that it contains neurones which respond to the sight of food, and which are influenced by learning. (Such pattern-specific visual responses, and their modification by learning, require forebrain areas such as the inferior temporal visual cortex and the amygdala, as described below.) This conclusion is consistent with the anatomy of the hypothalamus and substantia innominata, which receive projections from limbic structures such as the amygdala that in turn receive projections from the association cortex (Nauta 1961; Herzog and Van Hoesen 1976). The conclusion is also consistent with the evidence that decerebrate rats retain simple controls of feeding, but do not show normal learning about food (Grill and Norgren 1978). These rats accept sweet solutions placed in their mouths when hungry and reject them when satiated, so that some control of responses to gustatory stimuli which depend on hunger can occur caudal to the level of the hypothalamus. However, these rats are unable to feed themselves, and do not learn to avoid poisoned solutions. The importance of visual inputs and learning to feeding, in relation to which some hypothalamic neurones respond, is that animals, and especially primates, may eat many foods every day, and must be able to select foods from other visual stimuli, as well as produce appropriate preparative responses to them, such as salivation and the release of insulin.

FUNCTIONS OF THE TEMPORAL LOBE IN FEEDING

Bilateral damage to the temporal lobes of primates leads to the Kluver–Bucy syndrome, in which lesioned monkeys, for example, select and place in their mouths non-food as well as food items shown to them, and repeatedly fail to avoid noxious stimuli (Kluver and Bucy 1939; Jones and Mishkin 1972). Rats with lesions in the basolateral amygdala also display altered food selection, in that they ingest relatively novel foods (Rolls, E. T. and Rolls, B. J. 1973; Borsini and Rolls 1984), and fail to learn to avoid to ingest a solution which has previously resulted in sickness (Rolls, B. J. and Rolls, E. T. 1973). The basis for these alterations in food selection and in food-related learning are considered next.

The monkeys with temporal lobe damage have a visual discrimination deficit, in that they are impaired in learning to select one of two objects under which food is found, and thus fail to form correctly an association between the visual stimulus and reinforcement (Jones and Mishkin 1972; Spiegler and Mishkin 1981). This syndrome is produced by lesions which damage the cortical areas in the anterior part of the temporal lobe and the underlying amygdala (Jones and Mishkin 1972), or by lesions of the amygdala (Weiskrantz 1956; Spiegler and Mishkin 1981), or of the temporal lobe neocortex (Akert, Grueson, Woolsey, and Meyer 1961). Lesions to part of the temporal lobe neocortex, damaging the inferior temporal visual cortex and extending into the cortex in the ventral bank of the superior temporal sulcus, produce visual aspects of the syndrome, seen for example as a tendency to select non-food as well as food items (L. Weiskrantz, personal communication). Anatomically, there are connections from the inferior temporal visual cortex to the amygdala (Herzog and Van Hoesen 1976), which in turn projects to the hypothalamus (Nauta 1961), thus providing a route for visual information to reach the hypothalamus (Rolls 1981*b*). This evidence, together with the evidence that damage to the hypothalamus can disrupt feeding (Rolls 1981*b*; Winn, Tarbuck, and Dunnett 1984), indicates that there is a system which includes the visual cortex in the temporal lobe, projections to the amygdala, and further connections to structures such as the lateral hypothalamus, which is involved in behavioural responses made on the basis of learned associations between visual stimuli and reinforcers such as food. Given this evidence from lesion and anatomical studies, the contribution of each of these regions to the visual analysis and learning required for these functions in food selection will be considered using evidence from the activity of single neurones in these regions.

Recordings were made from single neurones in the inferior temporal visual cortex while rhesus monkeys performed visual discriminations, and while they were shown visual stimuli associated with positive reinforcement such as food, with negative reinforcement such as aversive hypertonic saline, and neutral visual stimuli (Rolls *et al*. 1977). It was found that during visual discriminations inferior temporal neurones often had sustained visual responses with latencies of 100–140 ms to the discriminanda, but that these responses did not depend on whether the visual stimuli were associated with reward or punishment (in that the neuronal responses did not alter during reversals, when the previously rewarded stimulus was made to signify aversive saline, and the previously punished stimulus was made to signify reward) (Rolls *et al*. 1977). The conclusion, that the responses of inferior temporal neurones during visual discriminations do not code for whether a visual stimulus is associated with reward or punishment, is also consistent with the findings of Ridley, Hester, and Ettlinger (1977), Jarvis and Mishkin (1977), Gross, Bender, and Gerstein (1979), and Sato Kawamura, and Iwai (1980). Further, it was found that inferior temporal neurones did not respond only to food-related visual stimuli, or only to aversive stimuli, and were not dependent on hunger, but rather that in many cases their responses depended on physical aspects of the stimuli such as shape, size, orientation, colour, or texture (Rolls *et al*. 1977).

These findings indicate that the responses of neurones in the inferior temporal visual cortex do not reflect the association of visual stimuli with reinforcers such as food. Given these findings, and the lesion evidence described above, it is thus likely that the inferior temporal cortex is an input stage for this process. The next structure on the basis of anatomical connections (see above) is the amygdala, and this will be considered next.

In recordings made from 1754 amygdaloid neurones, it was found that 113 (6.4 per cent) (of which many were in a dorsolateral region of the amygdala known to receive directly from the inferior temporal visual cortex (Herzog and Van Hoesen 1976)) had visual responses, which in most cases were sustained while the monkey looked at effective visual stimuli (Sanghera *et al*. 1979). The latency of the responses was 100–140 ms or more. The majority (85 per cent) of these visual neurones responded more strongly to some stimuli than to others, but physical factors which accounted for the responses such as orientation, colour and texture could not usually be identified. It was found that 22 (19.5 per cent) of these visual neurones responded primarily to foods and to objects associated with food, but for none of these neurones did the responses occur uniquely to food-related stimuli, in that they all responded to one or more aversive or neutral stimuli. Further, although some neurones responded in a visual discrimination to the visual stimulus which indicated food reward, but not to the visual stimulus associated with aversive saline, only minor modifications of the neuronal responses were obtained when the association of the stimuli with reinforcement was reversed in the reversal of the visual discrimination. Thus, even the responses of these neurones were not invariably associated with whichever stimulus was associated with reward. A comparable population of neurones with responses apparently partly, but not uniquely, related to aversive visual stimuli was also found (Sanghera *et al*. 1979).

These findings suggest that the amygdala could be involved at an early stage of the processing by which visual stimuli are associated with reinforcement, but that neuronal responses here do not code uniquely for whether or not a visual stimulus is associated with reinforcement. Neurones with responses more closely related to reinforcement are found in an area to which the amygdala projects, the lateral hypothalamus and substantia innominata, as decribed above. Thus, in the anatomical sequence inferior temporal visual cortex to amygdala to lateral hypothalamus and substantia innominata, there is evidence that neuronal responses become, through these stages, more relevant as a result of learning to the control of feeding, so that finally there are neurones in the lateral hypothalamus and substantia innominata with responses which occur only to stimuli which the organism has learned are food or signify food, and to which it is appropriate when hungry to initiate feeding.

FUNCTIONS OF THE ORBITOFRONTAL CORTEX IN FEEDING

Damage to the orbitofrontal cortex alters food preferences, in that monkeys with damage to this area select and eat foods which were previously rejected (Butter, McDonald, and Snyder 1969). Lesions of the orbitofrontal cortex also lead to a failure to correct feeding responses when these become inappropriate. Examples of the situations in which these abnormalities in feeding responses are found include (a) extinction, in that feeding responses continue to be made to the previously reinforced stimulus, (b) reversals of visual discriminations, in that the monkeys make responses to the previously reinforced stimulus or object, (c) Go/Nogo tasks, in that responses are made to the stimulus which is not associated with food reward, and (d) passive avoidance, in that feeding responses are made even when they are punished (Butter 1969; Iversen, and Mishkin 1970; Jones and Mishkin 1972; Tanaka 1973; see also Rosenkilde 1979; Fuster 1980). (It may be noted that, in contrast, the formation of associations between visual stimuli and reinforcement is much less affected by these lesions than by temporal lobe lesions, as tested during visual discrimination learning and reversals (Jones and Mishkin 1972)).

To investigate how the orbitofrontal cortex may be involved in feeding and in the correction of feeding responses when these become inappropriate, recordings were made of the activity of 494 orbitofrontal neurones during the performance of a Go/Nogo task, reversals of a visual discrimination task, extinction, and passive avoidance (Thorpe, Rolls, and Maddison 1983). First, neurones were found which responded in relation to the preparatory auditory or visual signal used before each trial (15.1 per cent), or non-discriminatively during the period in which the discriminative visual stimuli were shown (37.8 per cent). These neurones are not considered further here. Second, 8.6 per cent of neurones had responses which occurred discriminatively during the period in which the visual stimuli were shown. The majority of these neurones responded to whichever visual stimulus was associated with reward, in that the stimulus to which they responded changed during reversal. However, 6 of these neurons required a combination of a particular visual stimulus in the discrimination *and* reward in order to respond. Further, none of this second group of neurones responded to all the reward-related stimuli, including different foods which were shown, so that in general this group of neurones coded for a combination of one or several visual stimuli *and* reward. Thus information that particular visual stimuli had previously been associated with reinforcement was represented in the responses of orbitofrontal neurones. Third, 9.7 per cent of neurones had responses which occurred after the lick response was made in the task to obtain reward. Some of these responded independently of whether fruit juice reward was obtained, or aversive hypertonic saline was obtained in trials on which the monkey licked in error or was given saline in the first trials of a reversal. Through these neurones information that a lick had been made was represented in the orbitofrontal cortex. Other neurones in this third group responded only when fruit juice was obtained, and thus through these neurones information that reward had been given on that trial was represented in the orbitofrontal cortex. Other neurones in this group responded when saline was obtained when a response was made in error, or when saline was obtained on the first few trials of a reversal (but not in either case when saline was simply placed in the mouth), or when reward was not given in extinction, or when food was taken away instead of being given to the monkey. However, they did not respond in all these situations in which reinforcement was omitted or punishment was given. Thus through these neurones task-selective information that reward had been omitted or punishment given was represented in the orbitofrontal cortex.

These three groups of neurones found in the orbitofrontal cortex could together provide for computation of whether the reinforcement previously associated with a particular stimulus was still being obtained, and generation of a signal if a match was not obtained. This signal could be partly reflected in the responses of the last subset of neurones with task-selective responses to non-reward, or to unexpected punishment. This signal could be used to alter the monkey's behaviour, leading, for example, to reversal to one particular stimulus but not to other stimuli, to extinction to one stimulus but not to others, etc. It could also lead to the altered responses of the orbitofrontal differential neurones found as a result of learning in reversal, so that their responses indicate appropriately whether a particular stimulus is now associated with reinforcement.

Thus, the orbitofrontal cortex contains neurones which appear to be involved in altering behavioural responses when these responses are no longer associated with reward or become associated with punishment. In the context of feeding, it appears that without these neurones the primate is unable to suppress his behaviour correctly to non-food objects, in that altered food preferences are produced by orbitofrontal damage (Butter, McDonald, and Snyder 1969). It also appears that without these neurones the primate is unable to correct his behaviour when it becomes appropriate to break a learned association between a stimulus and a reward such as food (Jones and Mishkin 1972). The orbitofrontal neurones could be involved in the actual breaking of the association, or in the alteration of behaviour when other neurones signal that the connection is no longer appropriate. As shown here, the orbitofrontal cortex contains neurones with responses which could provide the information necessary for, and the basis for, the unlearning. This type of unlearning is important in enabling animals to alter the environmental stimuli to which motivational responses such as feeding have previously been made, when experience shows that such responses have become inappropriate. In this way they can ensure that their feeding and other motivational responses remain continually adapted to a changing environment.

FUNCTIONS OF THE STRIATUM IN FEEDING

Damage to the nigrostriatal bundle, which depletes the striatum of dopamine, produces aphagia and adipsia associated with a sensori-motor disturbance in the rat (Ungerstedt 1971; Marshall, Richardson, and Teitelbaum 1974; Stricker and Zigmond 1976; Stricker 1984). In order to analyse how striatal function is involved in feeding, the activity of single neurones is being recorded in different regions of the striatum during the initiation of feeding and during the performance of other tasks known to be affected by damage to particular regions of the striatum (Rolls 1979, 1984), as described below.

In the head of the caudate nucleus (Rolls, Thorpe, and Maddison 1983*c*), which receives inputs particularly from the prefrontal cortex, many neurones responded to environmental stimuli which were cues to the monkey to prepare for the possibile initiation of a feeding response. Thus, 22.4 per cent of neurones recorded responded during a cue given by the experimenter to the monkey that a food or non-food object was about to be shown, and fed if food. Comparably, in a visual discrimination task made to obtain food, 14.5 per cent of the neurones (including some of the above) responded during a 0.5 s tone/light cue which preceded and signalled the start of each trial. It is suggested that these neurones are involved in the utilization of environmental cues for the preparation for movement, and that disruption of the function of these neurones contributes to the akinesia or failure to initiate movements (including those required for feeding) found after depletion of dopamine in the striatum (Rolls *et al*. 1983*b*). Some other neurones (25.8 per cent) responded if food was shown to the monkey immediately prior to feeding by the experimenter, but the responses of these neurones typically did not occur in other situations in which food-related visual stimuli were shown, such as during the visual discrimination task. Comparably, some other neurones (24.3 per cent) responded differentially in the visual discrimination task, to for example the visual stimulus which indicated that the monkey could initiate a lick response to obtain food, yet typically did not respond when food was simply shown to the monkey prior to feeding. The responses of these neurones thus occur to particular stimuli which indicate that particular motor responses should be made, and they are therefore situation-specific, so that it is suggested that these neurones are involved in stimulus-motor response connections. In that their responses are situation-specific, they are different from the responses of the hypothalamic neurones with visual responses to the sight of food, described above (Rolls *et al*. 1983*c*). It is therefore suggested that these neurones in the head of the caudate nucleus could be involved in relatively fixed feeding responses to food made in particular, probably well-learned, situations, but do not provide a signal which reflects whether a visual stimulus is associated with food, and on the basis of which any response required to obtain the food could be initiated. Rather, it is likely that the systems in the temporal lobe, hypothalamus and orbitofrontal cortex described above, are involved in this more flexible decoding of the food value of visual stimuli.

In the tail of the caudate nucleus, which receives inputs from the inferior temporal visual cortex, neurones were found which responded to visual stimuli such as gratings and edges, but which showed habituation which was rapid and pattern-specific (Caan, Perrett, and Rolls 1984). It was suggested that these neurones are involved in orientation to patterned visual stimuli, and in pattern-specific habituation to these stimuli (Caan, Perrett, and Rolls 1984). These neurones would thus appear not to be involved directly in the control of feeding, although a disturbance in the ability to orient normally to a changed visual stimulus could indirectly have an effect on the ability to react normally.

In the putamen, which receives from the sensori-motor cortex, neurones were found with activity related to, for example, mouth or arm movements made by the monkey (Rolls, Thorpe, Boytim, Szabo, and Perrett 1984). Disturbances in the normal function of these neurones might be expected to affect the ability to initiate and execute movements, and therefore might indirectly affect the ability to feed normally.

Thus, in different regions of the striatum, neurones are found which may be involved in orientation to environmental stimuli, in the use of such stimuli in the preparation for and initiation of movements, in the execution of movements, and in stimulus-response connections appropriate for particular responses made in particular situations to particular stimuli. Disturbances of feeding produced by damage to the striatum might be expected to occur because of disruption of these functions, and not because the striatum plays a direct role in the regulation of food intake.

CONCLUSIONS

Investigations in non-human primates have provided evidence that the lateral hypothalamus and adjoining substantia innominata are involved in the control of feeding, because there is a population of neurones in these regions which responds to the sight and/or taste of food if the organism is hungry. The responses of these neurones may reflect the rewarding value or pleasantness of food, for stimulation in this region can mimic the reward value of food. It has been found that although after satiation with one food these neurones no longer respond to that food, they may still respond at least partly to other foods which have not been eaten. Following this finding, it has been shown that sensory-specific satiety is an important determinant of human food intake, and that associated with this, variety is an important factor in determining the amount of food eaten. A route for information about which visual stimuli are foods to reach the hypothalamus is provided by temporal lobe structures such as the inferior temporal visual cortex and amygdala, with the amygdala being important for learning which visual stimuli are foods. The orbitofrontal cortex contains a population of neurones that appear to be important in correcting feeding responses as a result of learning. The striatum contains neural systems which are important for the initiation of different types of motor and behavioural responses, including feeding.

References

Akert, K., Gruesen, R. A., Woolsey, C. N., and Meyer, D. R. (1961), Kluver-Bucy syndrome in monkeys with neocortical ablations of temporal lobe, *Brain* **84**, 480–98.

Anand, B. K. and Brobeck, J. R. (1951). Localization of a feeding centre in the hypothalamus of the rat. *Proc. Soc. Exp. Biol. Med.* **77**, 323–4.

Borsini, F. and Rolls, E. T. (1984). Role of noradrenaline and serotonin in the basolateral region of the amygdala in food preferences and learned taste aversions in the rat. *Physiol. Behav.* **33**, 37–43.

Burton, M. J., Rolls, E. T, and Mora, F. (1976). Effects of hunger on the responses of neurones in the lateral hypothalamus to the sight and taste of food. *Exp. Neurol.* **51**, 668–77.

Butter, C. M. (1969). Perseveration in extinction and in discrimination reversal tasks following selective prefrontal ablations in Macaca mulatta. *Physiol. Behav.* **4**, 163–71.

Butter, C. M., McDonald, J. A., and Snyder, D. R. (1969). Orality, preference behaviour, and reinforcement value of non-food objects in monkeys with orbital frontal lesions. *Science* **164**, 1306–7.

Caan, W., Perrett, D. I, and Rolls, E. T. (1983). Responses of striatal neurones in the behaving monkey. 2. Visual processing in the caudal neostriatum. *Brain Res.* **290**, 53–65.

Divac, I. (1975). Magnocellular nuclei of the basal forebrain project to neocortex, brain stem, and olfactory bulb. Review of some functional correlates. *Brain Res.* **93**, 385–98.

Friedman, M. I. and Stricker, E. (1976). The physiological psychology of hunger: a physiological perspective. *Psychol. Rev.* **83**, 409–31.

Fuster, J. M., (1980). *The prefrontal cortex,* Raven Press, New York.

Gallistel, C. R. and Beagley, G. (1971). Specificity of brain-stimulation reward in the rat. *J. Comp. Physiol. Psychol.* **76**, 199–205.

Gibbs, J., Maddison, S. P., and Rolls, E. T. (1981). The satiety role of the small intestine in sham feeding rhesus monkeys. *J. Comp. Physiol. Psychol.* **95**, 1003–15.

Grill, H. J. and Norgren, R. (1978). Chronically decerebrate rats demonstrate satiation but not bait shyness. *Science.* **201**, 267–9.

Gross, C. G., Bender, D. B, and Gerstein, G. L. (1979). Activity of inferior temporal neurones in behaving monkeys. *Neuropsychologia* **17**, 215–29.

Grossman, S. P. (1967). *A textbook of physiological psychology*. Wiley, New York.

Grossman, S. P. (1973). *Essentials of physiological psychology*. Wiley, New York.

Herzog, A. G. and Van Hoesen, G. W. (1976). Temporal neocortical afferent connections to the amygdala in the rhesus monkey. *Brain Res.* **115**, 57–69.

Hoebel, B. G. (1969). Feeding and self-stimulation. *Ann. N. Y. Acad. Sci.* **157**, 757-78.

Iversen, S. D. and Mishkin, M. (1970). Perseverative interference in monkey following selective lesions of the inferior prefrontal convexity. *Exp. Brain Res.* **11**, 376–86.

Jarvis, C. D. and Mishkin, M. (1977). Responses of cells in the inferior temporal cortex of monkeys during visual discrimination reversals. *Soc. Neurosci. Abstr.* **3**, 1794.

Jones, B. and Mishkin, M. (1972). Limbic lesions and the problem of stimulus-reinforcement associations. *Exp. Neurol.* **36**, 362–77.

Kievit, J. and Kuypers, H. G. J. M. (1975). Subcortical afferents to the frontal lobe in the rhesus monkey studied by means of retrograde horseradish peroxidase transport. *Brain Res.* **85**, 261–6.

Kluver, H. and Bucy, P. C. (1939). Preliminary analysis of functions of the temporal lobes in monkeys. *Arch. Neurol. Psychiatr.* **42**, 979–1000.

Le Magnen, J. (1980). The body energy regulation: the role of three brain responses to glucopenia. *Neurosci. Biobehav. Rev.* **4**, Suppl. 1, 65–72.

Marshall, J. F., Richardson, J. S, and Teitelbaum, P. (1974). Nigrostriatal bundle damage and the lateral hypothalamic syndrome. *J. Comp. Physiol. Psychol.* **87**, 808–30.

Mora, F., Rolls, E. T, and Burton, M. J. (1976). Modulation during learning of the responses of neurones in the hypothalamus to the sight of food. *Exptl. Neurol.* **53**, 508–19.

Nauta, W. J. H. (1961). Fibre degeneration following lesions of the amygdaloid complex in the monkey. *J. Anat.* **95**, 515–31.

Nauta, W. J. H. and Domesick, V. B. (1978). Crossroads of limbic and striatal circuitry: Hypothalamonigral connections. In *Limbic mechanisms* (eds. K. E. Livingstone and O. Hornykiewicz) pp. 75–93. Plenum Press, New York.

Olds, J. (1977). Drives and reinforcements: behavioural studies of hypothalamic functions. Raven Press, New York.

Ono, T., Nishino, H., Sasaki, K., Fukuda, M., and Muramoto, K. (1980). Role of the lateral hypothalamus and amygdala in feeding behaviour. *Brain Res. Bull.* **5**, Suppl. 4, 143–9.

Ridley, R. M., Hester, N.S., and Ettlinger, G. (1977). Stimulus- and response-dependent units from the occipital and temporal lobes of the unanaesthetized monkey performing learnt visual tasks. *Exp. Brain Res.* **27**, 539–52.

Rolls, E. T. (1975). *The brain and reward.* Pergamon Press, Oxford.

Rolls, E.T. (1976). The neurophysiological basis of brain-stimulation reward. In *Brain-stimulation reward* (eds. A. Wauquier and E. T. Rolls) pp. 65–87. Elsevier, North Holland, Amsterdam.

Rolls, E. T. (1979). Effects of electrical stimulation of the brain on behaviour. In *Psychology surveys* (ed. K. Connolly) Vol. 2, pp. 151–69. George Allen and Unwin, Hemel Hempstead, UK.

Rolls, E. T. (1981*a*). Processing beyond the inferior temporal visual cortex related to feeding, memory, and striatal function. In *Brain mechanisms of sensation* (eds. Y. Katsuki, R. Norgren, and M. Sato) Ch. 16, pp. 241–69. Wiley, New York.

Rolls, E. T. (1981*b*). Central nervous mechanisms related to feeding and appetite. *Br. Med. Bull.* **37**, 131–4.

Rolls, E. T. (1982). Feeding and reward. In *The neural basis of feeding and reward* (eds. B. G. Hoebel and D. Novin) pp. 323–7. Haer Institute for Electrophysiological Research, Brunswick, Maine.

Rolls, E. T. (1983). Feeding. In *Advances in vertebrate neuroethology* (eds. J. P. Ewert, R. R. Capranica, and D. J. Ingle) pp. 1067–86. Plenum Press, New York.

Rolls, E. T. (1984). Activity of neurones in the basal ganglia of the behaving monkey. In *The basal ganglia: structure and function* (eds. J. McKenzie and L. Wilcox). pp. 467–93. Plenum Press, New York.

Rolls, B. J. and Rolls, E. T. (1973*a*). Effects of lesions in the basolateral amygdala on fluid intake in the rat. *J. Comp. Phsysiol. Psychol.* **83**, 240–7.

Rolls, E. T. and Rolls, B. J. (1973*b*). Altered food preferences after lesions in the basolateral region of the amygdala in the rat. *J. Comp. Physiol. Psychol.* **83**, 248–59.

Rolls, E. T. and Rolls, B. J. (1977). Activity of neurones in sensory, hypothalamic, and motor areas during feeding in the monkey. In *Food intake and chemical senses* (eds. Y. Katsuki, M. Sato, S. F. Takagi, and Y. Oomura) pp. 525–49. University of Tokyo Press, Tokyo.

Rolls, E. T. and Rolls, B. J. (1982). Brain mechanisms involved in feeding. In *Psychobiology of human food selection* (ed. L. M. Barker) Ch. 3, pp. 33–62. AVI Publishing Co, Westport, Connecticut.

Rolls, E. T., Burton, M. J., and Mora, F. (1976). Hypothalamic neuronal responses associated with the sight of food. *Brain Res.* **111**, 53–66.

Rolls, E. T., Burton, M. J., and Mora, F. (1980). Neurophysiological analysis of brain-stimulation reward in the monkey. *Brain Res.* **194**, 339–57.

Rolls, E. T., Judge, S. J., and Sanghera, M. K. (1977). Activity of neurones in the inferotemporal cortex of the alert monkey. *Brain Res.* **130**, 229–38.

Rolls, E. T., Rolls, B. J., and Rowe, E. A. (1983*a*). Sensory-specific and motivation-specific satiety for the sight and taste of food and water in man. *Physiol Behav.* **30**, 185–92.

Rolls, B. J., Rowe, E. A., and Rolls, E. T. (1982). How sensory properties of foods affect human feeding behaviour. *Physiol. Behav.* **29**, 409–17.

Rolls, B. J., Rowe, E. A., Rolls, E. T., Kingston, B., Megson, A., and Gunary, R. (1981*a*). Variety in a meal enhances food intake in man. *Physiol. Behav.* **26**, 215–21.

Rolls, B. J., Rolls, E. T., Rowe, E. A., amd Sweeney, K. (1981*b*). Sensory specific satiety in man. *Physiol. Behav.* **27**, 137–42.

Rolls, E. T., Sanghera, M. K., and Roper-Hall, A. (1979). The latency of activation of neurones in the lateral hypothalamus and substantia innominata during feeding in the monkey. *Brain Res.* **164**, 121–35.

Rolls, E. T., Thorpe, S. J., Boytim, M., Szabo, I., and Perrett, D. I. (1984). Responses of striatal neurones in the behaving monkey. 3. Effects of iontophoretically applied dopamine on normal responsiveness. *Neuroscience.* **12**, 1201–12.

Rolls, E. T., Thorpe, S. J., and Maddison, S. P. (1983*c*). Responses of striatal neurones in the behaving monkey. 1. Head of the caudate nucleus. *Behav. Brain Res.* **7**, 179–210.

Rolls, B. J., Van Duijenvoorde, P. M., and Rowe, E. A. (1983*b*). Variety in the diet enhances intake in a meal and contributes to the development of obesity in the rat. *Physiol. Behav.* **31**, 21–7.

Rolls, B. J., Van Duijenvoorde, P. M., and Rolls, E. T. (1984). Pleasantness changes and food intake in a varied four course meal. *Appetite*. In press.

Rosenkilde, C. E. (1979). Functional heterogeneity of the prefrontal cortex in the monkey: review. *Behav. Neural Biol.* **25**, 301–45.

Sanghera, M. K., Rolls, E. T., and Roper-Hall, A. (1979). Visual responses of neurones in the dorsolateral amygdala of the alert monkey. *Exp. Neurol.* **63**, 610–26.

Saper, C. B., Loewy, A. D., Swanson, L. W., and Cowan, W. M. (1978). Direct hypothalamoautonomic connections. *Brain Res.* **117**, 305–12.

Saper, C. B., Swanson, L. W., and Cowan, W. M; (1979). An autoradiographic study of the efferent connections of the lateral hypothalamic area in the rat. *J. Comp. Neurol.* **183**, 689–706.

Sato, T., Kawamura, T., and E. Iwai (1980). Responsiveness of inferotemporal single units to visual pattern stimuli in monkeys performing discrimination. *Exp. Brain Res.* **38**, 313–9.

Spiegler, B. J. and Mishkin, M. (1981). Evidence for the sequential participation of inferior temporal cortex and amygdala in the acquisition of stimulus-reward associations. *Behav. Brain Res.* **3**, 303–17.

Stellar, E. (1954). The physiology of motivation. *Psychol. Rev.* **61**, 5–22.

Stricker, E. M. (1984). Brain monoamines and the control of food intake. *Int. J. Obes.* In press.

Stricker, E. M. and Zigmond, M. J. (1976). Recovery of function after damage to central catecholamine-containing neurones: a neurochemical model for the lateral hypothalamic syndrome. *Prog. Psychobiol. Physiol. Psychol.* **6**, 121–188.

Tanaka, D. (1973). Effects of selective prefrontal decortication on escape behaviour in the monkey. *Brain Res.* **53**, 161–73.

Thorpe, S. J., Rolls, E. T., and Maddison, S. (1983). Neuronal activity in the orbitofrontal cortex of the behaving monkey. *Exp. Brain Res.* **49**, 93–115.

Ungerstedt, U. (1971). Adipsia and aphagia after 6-hydroxydopamine induced degeneration of the nigrostriatal dopamine system. *Acta physiol. Scand.* **81**, Suppl. 367, 95–122.

Weiskrantz, L. (1956). Behavioural changes associated with ablation of the amygdaloid complex in monkeys. *J. Comp. Physiol. Psychol.* **49**, 381–91.

Winn, P., Tarbuck, A. and Dunnett, S. B. (1984). Ibotenic acid lesions of the lateral hypothalamus: comparison with electrolytic lesion syndrome. *Neuroscience,* **11**. In press.

Woods, S. C., McKay, L. D., Stein, L. J., West, D. B., and Porte, D. (1980). Neuroendocrine regulation of food intake and body weight. *Brain Res. Bull.* **5**, Suppl. 4, 1–5.

2

Neuropeptides and food and water intake

STEVEN J. COOPER

INTRODUCTION

Peptides are biologically active molecules which consist of sequences of amino-acid residues linked by peptide bonds. Neuropeptides constitute an extensive (and growing) family which have been identified within neurones of the central and peripheral nervous systems. They are synthesized by selective cleavage of the amino acid sequences of larger precursor molecules, stored in neurones and released from them. Together, they comprise the latest generation of molecules which are believed to act as important elements in the signalling systems employed by nerve cells of all varieties. They can be described, according to function, as neurotransmitters, neuromodulators, neurohormones or messengers (Ajmone-Marsan and Traczyk 1980; Barker and Smith 1980; Costa and Trabucchi 1980; Iversen, Iversen, and Snyder 1983).

This chapter reviews some of the recent evidence which implicates neuropeptides in the control of food and water intake. The sheer volume of relevant research findings precludes a comprehensive account being possible in the present context. I have had to be selective in my coverage, therefore, and have excluded from consideration in the first instance, two major topics, each of which encompasses a considerable literature. One topic, not considered here, is the important role of angiotensin in thirst and sodium appetite. Angiotensin is not only an extremely potent dipsogenic stimulus, but it also powerfully stimulates sodium appetite. The decision to exclude angiotensin can be justified, however, because excellent, comprehensive reviews which deal with these actions are available (Epstein 1982; Fitzsimons 1979, 1980; Rolls and Rolls 1982). The status of angiotensin as a neuropeptide in the central nervous system is documented in detail by Lang, Unger, Rascher, and Ganten (1983), and by Phillips (1980). Also excluded is consideration of the endogenous opioid peptides, the enkephalins and endorphins. The decision was taken reluctantly, because there are many interesting developments which implicate these peptides in the control of feeding and drinking responses (Cooper and Sanger 1984). Once more, however, excellent reviews of the rapidly-multiplying literature are available (Morley, Levine, Yim, and Lowy 1983; Sanger 1981, 1983). The involvement of endogenous opioid peptides in feeding and drinking responses has its complexities, but they may contribute to the instigation and maintenance of ingestional responses. This serves to distinguish them from the several neuropeptides which do figure in the following sections of the chapter. The rich variety of the endorphin family is graphically described by Bloom (1983), while detailed accounts are available of the brain enkephalin (Miller 1983), and endorphin (Akil and Watson 1983) systems.

Even after these exclusions, a current list of known neuropeptides remains impressive (Table 2.1). Authors are quick to observe that many more neuropeptides probably wait to be discovered (Cooper, Bloom, and Roth 1982; Iversen 1983). Some are already familiar from other contexts, as gut, pancreatic, or pituitary hormones. Some were first isolated within the central nervous system, only to be identified subsequently in peripheral nerves, and in non-neural tissues. It is a source of considerable fascination that neuropeptide molecules turn up not only in nerve cells of every kind, but also in endocrine and paracrine cells, in the retina, and in the skin. The ubiquitous distribution of neuropeptides renders an encompassing understanding of their functions, and of the mechanisms by which their effects are mediated, an exceedingly complex issue. This is particularly so when behavioural variables are the subject of investigation. Feeding and drinking responses, we know, are determined and influenced by many factors. Potentially, the number of different ways in which neuropeptides could relate to the determinants and mediators of ingestional responses are legion. It is true to say that, in our present state of knowledge, there is no neuropeptide for which we have more than rudimentary information of its possible involvement in the controls of feeding and/or drinking behaviour.

This chapter deals, first, with a selected group of neuropeptides, each of which has been proposed to act as a satiety signal. As feeding is brought to an end, and a condition of satiety ensues, it has been suggested that neuropeptides are instumental in terminating the meal. The ingestion of food may bring about the release of endogenous neuropeptides, which in turn may signal satiety. We shall begin with cholecystokinin, for which there is now a considerable literature.

TABLE 2.1 *Neuropeptides which are found within the mammalian CNS*

ACTH	α-MSH
Angiotensin II	Motilin
Avian pancreatic polypeptide	Neuropeptide Y
Bombesin	Neurotensin
Bradykinin	Oxytocin
Carnosine	Pancreatic polypeptide
Cholecystokinin (CCK)	Proctolin
Corticotropin-releasing factor (CRF)	Secretin
Gastrin	Somatostatin
Glucagon	Substance P
Growth hormone	Thyrotropin-releasing hormone (TRH)
Insulin	Vasopressin
β-Lipotropin	Vasoactive intestinal polypeptide (VIP)
Luteinizing-hormone-releasing hormone (LH-RH)	

This list excludes opioid peptides (enkephalins and endorphins). See Bloom (1983)

Known to be a gut hormone, it was the first peptide to be put forward as a putative satiety signal. More candidates have followed. This chapter therefore goes on to consider bombesin, glucagon, somatostatin and others. For reference, and for comparison with the first table, Table 2.2 lists some of the known gut peptides. A major research problem involves assessing the extent to which peripherally-located neuropeptides (for example, present in the gut) and/or centrally-situated neuropeptides participate in the business of signalling the termination of meals.

The chapter concludes with an examination of the intriguing effects of a group of neuropeptides, which were first isolated in the skin of amphibians. For some reason, not yet explained, these substances are potent dipsogens in birds, and yet are markedly antidipsogenic in the rat.

CHOLECYSTOKININ

Isolation

The isolation, structural identification and many of the peripheral physiological and pharmacological actions of cholecystokinin have been described in detail by Mutt (1980). Ivy and Oldberg (1928) provided evidence for the presence of a substance in the upper intestine of cats and dogs which caused the gallbladder to contract. They named the putative hormone, cholecystokinin. Harper and Raper (1943) subsequently showed that a substance obtained from extracts of porcine intestinal tissue was able to stimulate the release of pancreatic enzymes. They named this substance, pancreozymin. It was some time later that a substance which exhibited properties of both cholecystokinin and pancreozymin was isolated from porcine intestinal tissue by Jorpes and Mutt (1966).

The cholecystokinin (CCK) identified in Mutt's laboratory was shown to have 33 amino acid residues (CCK-33). The sequence of amino acids is shown in

TABLE 2.2 *Some peptides present in the gut*

Cholecystokinin: CCK 39	Neurotensin
CCK 33	Pancreatic polypeptide
CCK 8	Somatostatin
Enkephalins: Leu	Secretin
Met	Substance P
Gastric inhibitory peptide (GIP)	Thyrotropin-releasing hormone (TRH)
Gastrin releasing peptide (GRP)	Vasoactive intestinal peptide (VIP)
Bombesin	
Gastrin: G34	
G17	
G14	
Glucagon	
Motilin	

Amino-acid sequences of these peptides are provided in Walsh (1981)

Table 2.3, together with the sequences of related peptides. A variant with 39 amino acid residues was also isolated (CCK-39). A shorter fragment, cholecystokinin octapeptide (CCK-8) have been found to occur in free form in intestinal tissue of various species. It is worth noting that the C-terminal pentapeptide sequence of the cholecystokinins is the same as the corresponding sequence in the gastrins and in caerulein and phyllocaerulein (Table 2.3).

Distribution

Cells which synthesize or store CCK have been identified in the intestinal mucosa (Buffa, Solcia, and Go 1976; Polak, Pearse, Bloom, Buchan, Rayford, and Thompson 1975). In several mammalian species, these cells are most numerous in the duodenum and proximal jejunum, but absent from the distal ileum. Cholecystokinin is present not only in intestinal endocrine and paracrine cells, but also in gut nerves. Cholecystokinin-octapeptide-like immunoreactivity has been identified in the vagus nerve of the dog, and in enteric nerve fibres distributed in the enteric plexuses and also in mucosa (Dockray, Vaillant, and Hutchinson 1981).

Cholecystokinin is also found in the brain and pituitary (Beinfeld 1983; Emson and Marley 1983). CCK peptides appear to exist in multiple forms, although the bulk of the CCK-like material in mammalian brain resembles CCK-8 sulphate. Cholecystokinin is the most abundant neuropeptide so far discovered in the brain. In parts of the rat cortex, for example, the CCK content approaches that of the catecholamines.

Cholecystokinin is widely distributed throughout the brain, but is particularly concentrated in forebrain structures (cerebral cortex, amygdala, hippocampus,

TABLE 2.3 *Amino-acid sequences of cholecystokinin-like peptides*

CCK–33	NH_2-Lys-Ala-Pro-Ser-Gly-Arg-Val-Ser-Met-Ile-Lys-Asn-Leu-Gln-Ser-Leu-Asp-Pro-Ser-His-Arg-Ile-Ser-Asp-Arg-Asp-Tyr(SO_3)-Met-Gly-Trp-Met-Asp-Phe-ONH_2
CCK–8	NH_2-Asp-Tyr(SO_3)-Met-Gly-Trp-Met-Asp-Phe-ONH_2
Caerulein	Pyr-Gln-Asp-Tyr(SO_3)-Thr-Gly-Trp-Met-Asp-Phe-ONH_2
Phyllocaerulein	Pyr-Glu-Tyr(SO_3)-Thr-Gly-Trp-Met-Asp-Phe-ONH_2
Gastrin–17 (porcine)	Pyr-Gly-Pro-Trp-Met-Glu-Glu-Glu-Glu-Glu-Ala-Tyr(SO_3)-Gly-Trp-Met-Asp-Phe-ONH_2

and hypothalamus). Results for porcine, guinea pig, bovine, rhesus monkey, and human brain appear to be in fair agreement with the distribution of CCK reported for rat brain (Beinfeld 1983; Emson and Marley 1983). An interesting feature concerning CCK is that it is found to co-exist in neurones with dopamine (Hökfelt, Rehfeld, Skirboll, Ivemark, Goldstein, and Markey 1980), substance P (Skirboll, Hökfelt, Rehfeld, Cuello, and Dockray 1982), oxytocin (Vanderhaeghen, Lotstra, Vandersande, and Dierickx 1981), or with enkephalins (Martin, Geis, Holl, Schäfer and Voigt 1983).

Receptors

Using [^{125}I]-labelled CCK-33, high-affinity, specific binding sites for CCK have been discovered in rat (Hays, Beinfeld, Jensen, Goodwin, and Paul 1980; Saito, Sankaran, Goldfine, and Williams 1980), mouse (Saito, Goldfine, and Williams 1981), guinea pig (Zarbin, Innis, Wamsley, Snyder, and Kuhar 1981), and human brain (Hyas, Goodwin, and Paul 1981). The distribution of CCK binding sites in rat brain is fairly well correlated with CCK content. In the guinea pig brain, however, the highest concentration of binding sites is found in the cerebellum, which contains very little CCK.

Peripheral effects

In addition to causing the gallbladder to contract, CCK also stimulates the release of the pancreatic hormones, insulin and glucagon, inhibits gastric emptying, stimulates the secretion of hepatic bile, shortens intestinal transit time, and stimulates gastric-acid secretion (Mutt 1980).

Central effects

Effects of CCK which are thought to be mediated centrally include the inhibition of exploratory rearing, catalepsy, analgesia, ptosis, anticonvulsant activity and a potentiation of hexobarbital-induced sleeping in mice (Zetler 1980, 1981).

Food intake

(a) *Satiety signal-introduction*

Cholecystokinin has been proposed as a satiety signal, which is released as a consequence of ingesting food and which acts to bring feeding to an end. In one form of the hypothesis the entry of food into the gut stimulates release of endogenous CCK and this, by some means or other conveys a signal to the brain to promote satiety. Cholecystokinin, in this case, is proposed to function physiologically as a gut hormone, the stimulus for its release being dependent on food ingestion.

This important hypothesis rests on observations that exogenous CCK, administered either as an impure preparation of CCK-33 or as synthetic CCK-8, suppresses food consumption in animal and human subjects. A great deal of research has been generated in attempts to confirm the observations, assess the degree of behavioural specificity, and to determine the possible mechanisms by which exogenous CCK is effective in reducing food consumption. Some of the relevant literature has been reviewed previously (Morley 1982; Mueller and Hsiao 1978; Rehfeld 1980; Smith and Gibbs 1979).

The discovery of CCK systems within the central nervous system, and elsewhere, in addition to the CCK located in the gut, considerably complicates the experimental problem of determining the role(s) of endogenous CCK in relation to feeding termination. In addition, as the following sections show, CCK is only the first of a number of neuropeptides, which are present in neural and endocrine cells in the brain, gut and elsewhere, to be regarded as candidate satiety signals. Assessing the contributions of each of these substances and discovering the nature of possible interrelationships between them, relative to the control of feeding, is clearly a formidable task.

The literature devoted to CCK and food consumption is currently by far the largest for any neuropeptide thought to be implicated in the control of feeding responses. Cholecystokinin can serve, therefore, as a model for other putative satiety factors. The state of our knowledge concerning each of the other candidate peptides can be compared with what has been established in the case of CCK. It is important to stress, at the outset, that the release of *endogenous* CCK, acting physiologically to promote feeding satiety, has not been demonstrated. The effects of endogenous CCK release are inferred largely from studies in which exogenously-administered CCK has been used, supplemented by evidence from the effects of CCK receptor antagonists, and of substances which stimulate the release of endogenous CCK.

Gibbs, Young, and Smith (1973*a*) were the first to report that the administration of exogenous CCK significantly reduced food consumption in rats. Mildly food-deprived animals were injected intraperitoneally with partially purified (10 per cent) porcine CCK in doses of 2.5–40 Ivy dog units/kg body weight*. The extract, containing CCK, inhibited food intake. The lowest effective dose was found to be 5 IDU/kg, and the suppressant effect was dose-related. The effect of CCK was transient, and was limited to the first 30 min access to the food. Similar effects on food intake were obtained using synthetic CCK-8 and the structurally-related peptide, caerulein (extracted from the skin of the frog *Hyla caerulea*).

Gibbs *et al.* (1973*a*) performed a number of control experiments to assess the nature of the feeding-suppressant effect of CCK. Thus, they showed that CCK also reduced the consumption of a liquid food in 17-h deprived rats, which ruled out the possibility that CCK was merely interfering with the motor acts involved in the consumption of solid food. Water consumption in 12-h water-deprived rats was unaffected by CCK, which excluded the possibility that the peptide simply depressed behaviour non-specifically. Inhibition of food intake could arise if CCK made the animals sick, or if CCK administration had aversive consequences. However, the authors noted that rats ate rapidly initially, and only stopped eating sooner, following an injection of even a large dose of CCK.

* The Ivy dog unit (IDU) is defined in Ivy and Janacek (1959).

Furthermore, they did not obtain a conditioned taste aversion when CCK administration (40 IDU/kg) was paired with a novel sodium saccharin solution. Gibbs *et al.* (1973*a*) concluded that CCK may play an inhibitory role in the short term control of feeding behaviour.

(b) *Phylogenetic comparisons*

Cholecystokinin has been shown to suppress food consumption in mammalian species other than rats, and also in birds. Thus, CCK reduced food intake in fasted genetically obese mice (ob/ob), in doses which did not affect drinking or rearing rates (Batt 1983; McLaughlin and Baile 1981; Parrott and Batt 1980). There are several reports that CCK attenuated food consumption in pigs (Anika, Houpt and Houpt 1981; Baldwin, Cooper, and Parrott 1983; Houpt 1983; Parrott and Baldwin 1981). A similiar suppression of food intake was obtained with caerulein (Anika *et al.* 1981). Cholecystokinin and caerulein also suppress food consumption in rabbits adapted to a food deprivation schedule (Houpt, Anika, and Wolff 1978). Food intake in fasted sheep was reduced when CCK-8 was continuously infused into the cerebral ventricles of sheep (Della-Fera and Baile 1979, 1980). Intravenous injections of CCK-8 and caerulein suppressed feeding in domestic fowls in a dose-related way (Savory and Gentle 1980). Intracerebroventricular infusion of CCK-8 also reduced feeding responses in chicks (Denbow and Myers 1982). Partially purified porcine CCK (20 per cent) and CCK-8 produced rapid, dose-related reductions in food consumption in food-deprived rhesus monkeys (Gibbs, Falasco, and McHugh 1976). In wolves, however, CCK failed to suppress food consumption (Morley, Levine, Plotka, and Seal 1983).

There are several reports concerning the effects of CCK on food consumption and appetite in human subjects. In one study, carried out using 12 non-obese men, intravenous infusion of CCK-8 (4 ng/kg/min) significantly reduced the intake of a liquified blend of several foods served as a luncheon (Kissileff, Pi-Sunyer, Thornton, and Smith 1981). The subjects ate less, and stopped eating sooner, but the initial rate of eating was unaffected. Some subjects reported a sick sensation, but this was not consistently related to the suppression of feeding. The authors raised the possibility that CCK-8 may be a potential therapeutic agent for weight control. In a subsequent study, they demonstrated that CCK-8, infused at the same rate, reduced food consumption in 6 of 8 obese subjects (weighing 137 per cent of average desirable weight), without producing overt side effects (Pi-Sunyer, Kissileff, Thornton, and Smith 1982). As in the first study, subjects stopped eating sooner, and CCK-8 did not change the rate of eating. The authors concluded that CCK-8 can serve as a short-term satiety signal in human subjects, and that the suppression of feeding was not due to sickness or abdominal discomfort.

In another study, 16 non-obese subjects were infused under double-blind conditions with saline or with various doses of 95 per cent pure porcine CCK-33 (0.6–6.0 IDU), following overnight fast (Stacher, Bauer, and Steinringer 1979). As the infusions were made, food was prepared and cooked in the presence of the subjects in order to arouse their appetite. The CCK infusions reduced the subjects' rating of feelings of hunger or voraciousness, even in the case of the smallest administered dose. Thus, CCK not only reduces food intake in human subjects, but also depresses reports of appetite aroused by the sight, sound, and smell of food preparation.

In earlier experiments, findings contradictory to these have been reported. Thus, Greenway and Bray (1977) failed to obtain any effect on feeding with intravenous or subcutaneous injections of CCK-8 administered to 14 normal subjects. Sturdevant and Goetz (1976) showed that the injection of partially purified CCK (20 per cent) decreased the consumption of a liquid meal in 10 non-obese subjects; in contrast, CCK infusions actually enhanced food intake.

The important factors which are responsible for the conflicting results are not known. Possibly, the slow infusion of CCK octapeptide was a significant element in the suppression of food intake in lean and obese subjects, observed by Kissileff and his colleagues. However, there remain many procedural differences amongst these available studies, and so the critical requirements for obtaining reliable suppression of food intake in humans, without adverse side effects, remain unidentified.

(c) *Ontogenetic comparisons*

Relatively little work has been done on the issue of the stage in development at which the feeding-suppressant action of cholecystokinin develops. Significant suppression of either solid or liquid diet consumption following intraperitoneal injection of 40 IDU/kg CCK-8 has been demonstrated in 21 day-old weanling rats (Bernstein, Lotter, and Zimmerman 1976). Water consumption in thirsty weanling rats was unaffected by CCK administration. Houpt and Houpt (1979) subsequently showed that intraperitoneal administration of 80 IDU/kg CCK-8 depressed milk ingestion (measured indirectly in terms of weight gain) in 3–7 day-old suckling rats. In adult rats, gastric loads of L-phenylalanine, but not of D-phenylalanine, release endogenous CCK (Meyer and Grossman 1972) and depress food intake (Anika, Houpt, and Houpt 1977). In the suckling rats, however, it seems that L-phenylalanine may fail to release CCK (Houpt and Houpt, 1979). Although the rats were responsive to exogenous CCK, Houpt and Houpt (1979) suggest that the controls of gastric CCK secretion had not matured in the suckling animals. If they are correct, we do not yet know the stage at which controls do mature.

(d) *Obesity*

Recently, Pi-Sunyer *et al.* (1982) reported that slow intravenous infusions of CCK-8 significantly reduced food consumption in obese human subjects.

Cholecystokinin octapeptide suppressed feeding in rats which had sustained lesions of the ventromedial nucleus of the hypothalamus (VMH) (Kulkosky, Breckenridge, Krinsky, and Woods 1976). VMH-lesioned rats are hyperphagic and become obese (Brobeck, Tepperman, and Long 1943). Evidently, the integrity of the VMH region is not a necessary condition for CCK-8 to elicit its satiety effect.

There is an accumulation of evidence which suggests that genetically obese animals differ in their sensitivity to CCK's effects, compared with lean controls. The adult Zucker obese rat, for example, appears to be less sensitive to the food suppressant effect of CCK-8 than lean control rats (McLaughlin and Baile 1979, 1980*b*). Compared with lean rats, Zucker fatty rats are hyperphagic and eat bigger meals. They consume about the same amount of food during the nocturnal period, but exhibit a hyperphagic feeding pattern during the day (Becker and Grinker 1977). Zucker obese rats were less sensitive to the suppressant effect of CCK-8 on feeding during the light period (McLaughlin and Baile 1980*b*). In addition, weanling Zucker obese rats were shown to be less sensitive to CCK's feeding-suppressant effect (McLaughlin and Baile 1980*a*). Zucker obese rats also appear to be less sensitive than lean rats to the effects of exogenously administered CCK on pancreatic exocrine function (McLaughlin, Peikin, and Baile 1982).

Apparently, Zucker obese rats are not invariably less sensitive to the effects of exogenously administered CCK-8. In fact they appear to be more sensitive than lean controls to the effects of the peptide on gastrointestinal responses (Moos, McLaughlin, and Baile 1982). However, if Zucker obese rats show a greater sensitivity to CCK's effects on feeding responses, they should show a similiar greater sensitivity in response to the modulation of secretion of endogenous cholecystokinin. When the secretion of CCK was increased by the administration of L-phenylalanine or of aprotinin (a trypsin inhibitor), food intake was decreased in Zucker obese and lean rats, but under these conditions, the obese rats were affected more than were lean animals (McLaughlin, Peikin, and Baile 1983*a*). Thus, the feeding response of obese rats to the manipulation of endogenous release of CCK contrast with the decreased feeding shown by these animals to exogenously administered CCK-8. These studies remain difficult to interpret. Zucker obese rats differ from lean controls in more ways than patterns of food intake. Interactions of CCK-8 with variables other than food consumption may play a prominent part in the outcome of these experiments. It is difficult to adjust CCK doses to achieve matching between obese and lean animals, and therefore it may not be clear to what degree differential sensitivity in response is due to different plasma CCK levels. But the contradictory nature of the results themselves, when the effects of exogenous CCK and manipulation of endogenous CCK secretion are considered, creates difficulties in interpretation. The significance of the results for a putative satiety function of CCK are difficult to assess.

There is a report that obese mice (ob/ob) are less sensitive to the effects of CCK-8 on food consumption compared with lean mice (McLaughlin and Baile 1981). Opposing this report, however, there is evidence for enhanced responsiveness of obese mice to CCK compared with their wild-type littermates (Batt 1983; Parrott and Batt 1980).

It is certain that obese animals and humans do display reduced food intake following the administration of CCK. However, the evidence available to date is not clear on whether or not obese animals are affected in terms of food consumption by CCK administration or release, in ways which are importantly different from the manner in which lean animals are affected.

(e) *Behavioural and electroencephalographic characteristics*

Cholecystokinin reduces food consumption in hungry animals of many specie and in sham-feeding animals (see Section F, below). The effectiveness of CC in rats to suppress food intake seems to be independent of the level of foo deprivation (Mueller and Hsiao 1979). Cholecystokinin reduces food ingestion i non-deprived animals, in which feeding has been aroused by tail-pinch (Nemerof Osbahr, Bissette, Jahnke, Lipton, and Prange 1978), or by food-associate external cues (Schallert, Pendergrass, and Farrar 1982).

When CCK affects food consumption, typically it does not affect the initi rate of eating within a meal. Instead, CCK acts to terminate feeding soon and so reduce meal size (Gibbs *et al.* 1973*a*; Hsiao, Wang, and Schallert 197 Kissileff *et al.* 1981; Pi-Sunyer *et al.* 1982). The inter-meal interval, that is th time to the subsequent meal, may be extended (Hsiao *et al.* 1979). These effec are what would be expected from a satiety signal. They are distinguishable, fo example, from the effect of amphetamine which suppresses food intake by delay in the initiation of a meal (Blundell and Leshem 1975).

In the rat, and other species, satiety can be identified by the presence of characteristic sequence of behaviours. The rat ceases feeding, engages in som non-feeding activities (grooming, movement about the cage), and then rests c falls asleep (Antin, Gibbs, Holt, Young, and Smith 1975; Blundell 1981; Boll 1960; Richter 1922; Smith and Gibbs 1979). A putative short-term satiety sign should not only contribute to the cessation of feeding; its effect should also b associated with the occurrence of the behavioural satiety sequence (BSS), an with the absence of abnormal behaviour patterns.

Antin *et al.* (1975) showed that injection of 10 per cent partially purifie porcine CCK (40 IDU/kg, intraperitoneal injection) brought feeding to a halt i sham-feeding rats, and elicited the typical BSS. In contrast to the effects c CCK, adulteration of the liquid diet with quinine stopped feeding in these rat but did not elicit the BSS. Instead the animals were very active and failed t rest. Clearly, CCK did not render the food aversive, and an end to feeding is nc a sufficient condition for the BSS to occur.

In addition to its behavioural features, postprandial satiety is also identifiab by characteristic changes in quantifiable electroencephalographic activity, fro an activated state towards synchronous, high voltage, low frequency wave Mansbach and Lorenz (1983) reported that fasted rats given CCK-8 (5–8 IDU/kg, intraperitoneally), in association with a meal, showed suppression c food intake, and electroencephalographic activity characteristic of rest and slee The effects they describe were dose-related. There was a progression fro resting, through synchronous wave sleep, to desynchronous wave sleep, wit increasing CCK dose level. The behavioural and electroencephalographic patter of rats following 80 IDU/kg were very similar to the patterns exhibited by pos prandial control rats.

(f) *Sham feeding*

Feeding, without satiety, can be induced in rats by providing 17 h food deprive animals with a liquid diet, and allowing the consumed food to drain from th

:omach by way of a surgically-implanted gastric fistula, which emerges from ıe abdominal wall. Under these circumstances, with the gastric fistula, animals ɔnsume the diet continuously and do not display signs of satiety (Antin *et al.* 975; Young, Gibbs, Antin, Holt, and Smith 1974). When the fistula is closed, nd the food accumulates in the stomach and enters the duodenum normally, ıe animals do satiate. The sham-feeding rat is a preparation which allows the ɔntribution of food accumulation in the stomach, and post-gastric effects of ɔod ingestion to be assessed in relation to the actions of putative satiety signals.

Intraperitoneal injection of partially purified CCK, or CCK-8, administered to ıe sham-feeding rat with an open gastric fistula, brings liquid diet consumption) an end, and elicits a normal BSS, without producing any evident signs of lness or distress (Antin *et al.* 1975; Gibbs, Young, and Smith 1973*b*). Lack of ɔod accumulation in the stomach, or lack of entry of food into the small ıtestine does not prevent the satiety effect of CCK occurring.

The effects of CCK on sham-feeding in the rat are paralleled by those of uodenal infusions of liquid diet (Liebling, Eisner, Gibbs, and Smith 1975). uodenal infusion of diet elicited feeding-cessation and the behavioural equence associated with satiety (BSS), but did not serve as an unconditioned :imulus for the formation of a conditioned aversion to saccharin flavour. eibling *et al.* (1975) referred to the effect of duodenal infusion in terms of ntestinal satiety', and speculated that it might be a condition which leads to ıe release of endogenous gut CCK.

Antin, Gibbs, and Smith (1977) proceeded to show that pregastric or gastric :imulation is an important requirement for the expression of intestinal satiety roduced by duodenal infusion of liquid diet. The infusion alone did not ınction as a satiety stimulus. What was required for feeding to satiate was the ɔncurrent ingestion of food with duodenal infusion of food. Evidently a combination of events involving entry of food to the stomach, on the one hand, and) the duodenum, on the other, is necessary for satiety to be fully expressed. A elated experiment showed that, in a comparable manner, CCK functions as a ıtiety signal in the sham-feeding rat, provided that feeding has begun, and food passing to the stomach (Antin, Gibbs, and Smith 1978). When CCK (40)U/kg, 20 per cent partially purified porcine CCK) was injected 12 min before ıe onset of sham-feeding, little suppressant effect on feeding was subsequently bserved. Sham-feeding was suppressed, however, when CCK was administered 2 min after sham-feeding had begun. Hence, it was combination of sham-eeding together with CCK administration which evoked satiation in the animals.

The satiety effect of CCK in the sham-feeding rat was not replicated by three ther gut hormones, gastrin, secretin or gastric inhibitory polypeptide (GIP) .orenz, Kreielsheimer, and Smith 1979). These data indicate a degree of structural specificity underlying CCK's satiety effect, and can be used to exclude ɔssible mechanisms by which CCK may achieve its satiety effect. For example, nce gastrin and CCK are equipotent in terms of acid secretion in the rat (Chey, ivasomboon, and Hendricks 1973), this shared action of theirs is unlikely to ɔntribute to CCK's satiety effect.

When rats with a gastric fistula sham feed, following only 3-h food deprivation, they do exhibit satiety and it is possible to measure meal size, latency to postprandial rest, and the interval to the next meal. Kraly, Carty, Resmick, and Smith (1978) were able to show that intraperitoneal administration of CCK to sham-feeding rats made their pattern of behaviour conform closely to control conditions, in which rats fed with the gastric fistula closed. Thus, CCK decreased meal size, decreased the latency to rest and increased the inter-meal interval in sham-feeding rats. The effect was specific to feeding, since CCK administration did not affect water consumption in 4 h water-deprived rats.

Cholecystokinin has been reported to suppress sham-feeding in rhesus monkeys (Falasco, Smith, and Gibbs 1979).

(g) *Gastric emptying*

In dogs, physiologically-relevant doses of CCK inhibit the rate of gastric emptying (Debas, Farrooq, and Grossman 1975). A mechanism by which CCK acts as a satiety signal could involve this inhibitory effect on gastric emptying. Moran and McHugh (1982) reported that, in rhesus monkeys, CCK-8 inhibited gastric emptying, and also suppressed food intake, provided CCK was given in combination with a test meal of 150 ml. 0.9 per cent saline infused into the stomach. They argued that entry of food into the intestine elicits CCK release, which then acts to delay gastric emptying and so allows some gastric distension to occur as food accumulates in the stomach. Feeding remains unaffected until a critical level of distension is reached. In the absence of food in the stomach, according to their proposal, CCK will remain ineffective. Only when food is available, and some gastric emptying occurs, can the effect of CCK be detected. This suggestion could be consistent with the sham-feeding data, described above, which show that CCK is an effective satiety signal only when the rats have started to ingest food which passes into the stomach.

In suckling rats, however (Houpt and Houpt 1979), a dose of CCK-8 which was effective in depressing intake did not slow gastric emptying. In these young animals, gastric emptying rate may not be a determinant of feeding satiety. These results do not exclude the possibility that in more mature rats, endogenous CCK release and retarded gastric emptying do play some part in satiation processes.

(h) *Vagotomy*

Several experiments have been undertaken to determine whether or not the vagus nerve is necessary to the satiety effect of CCK. Anika *et al.* (1977) reported that 80 IDU/kg CCK-8 injected intraperitoneally did suppress food consumption in 15 h food-deprived vagotomized rats. They considered that the intact vagus was not required for CCK's satiety effect. In contrast there are reports that abdominal vagotomy did abolish the suppressant effect of CCK on food intake (Lorenz and Goldman 1982; Morley, Levine, Kneip, and Grace 1982*b*; Smith, Jerome, Cushin, Eterno, and Simansky 1981). It is not clear why the discrepancy in results occurred, since doses of CCK-8 used in the three studies were comparable. Possibly, differences in the vagotomy procedure may have been responsible. Smith *et al.* (1981), indeed, showed that selective transection of hepatic, coeliac, or hepatic plus coeliac branches of the vagus had no effect on CCK's satiety effect. Transection of the gastric branch, however, was sufficient to abolish CCK's effect. Hence, vagal innervation of the stomach (and other structures supplied by the gastric branch-first part of the duodenum, liver, pancreas) appears to be involved in an important way in the short term satiety effect of exogenously-administered CCK.

(i) *Insulin*

It has been reported that CCK-8 stimulates insulin release (Frame, Davidson, and Sturdevant 1975), while steady intraperitoneal infusion of insulin has been shown to decrease food intake (VanderWeele, Pi-Sunyer, Novin, and Bush 1980). Perhaps there is an interaction between CCK and insulin to produce short-term postprandial satiety. However, VanderWeele (1982) failed to detect an interaction, and so concluded that there was no dependence between CCK and the release or availability of insulin producing satiety.

(j) *Peripheral site(s) of action*

The vagotomy experiment of Smith *et al.* (1981) indicated that structures innervated by the gastric branch of the vagus may be involved in the satiety signal yielded by CCK's action. It is unlikely, however, that the stomach is the site of CCK action, since the infusion of CCK-8 into the arterial supply of the stomach in pigs produced a slight but consistent increase in food consumption (Houpt 1983). The failure of intraportal infusions of CCK-8 to affect feeding in these animals suggested that CCK does not act upon receptors in the liver to produce satiation. Houpt's (1983) results suggested that tissues supplied by the coeliac and mesenteric arteries may act as sites for the CCK action to suppress food intake. These tissues include the small intestine, spleen, pancreas, and gall bladder, although the stomach can be excluded from consideration in the light of CCK's effect when infused into the stomach's arterial supply. Additional work will be required in order to specify the sites at which the satiety effects of peripherally-administered CCK are mediated.

(k) *Pharmacology*

Feeding can be induced by intracerebroventricular (icv) administration of noradrenaline, dynorphin or muscimol. In a series of experiments, Morley, Levine, and their colleagues have shown that subcutaneously-administered CCK-8 attenuated the feeding induced by noradrenaline (20 μg icv), but did not affect the feeding induced by dynorphin- (1-13) (10 μg icv) or by the GABA agonist, muscimol (100 ng icv) (Morley, Levine, and Kneip 1981; Morley, Levine, Grace, and Kneip 1982*a*; Morley, Levine, Murray, and Kneip 1982*c*). Cholecystokinin octapeptide also suppressed the feeding induced by subcutaneously administered insulin (Levine and Morley 1981). McCaleb and Myers (1980) showed that intraperitoneal infusions of CCK blocked the rat's feeding response to infusion of noradrenaline into diencephalic sites.

An interaction between noradrenergic processes and CCK is also suggested by evidence that clonidine (an α_2 receptor agonist) potentiated the feeding suppressant effect of CCK-8, while phenoxybenzamine, tolazoline, and yohimbine antagonized CCK's effects (Wilson, Denson, Bedford, and Hunsinger 1983). Furthermore morphine, haloperidol or picrotoxin also reduced the effectiveness of CCK-8, but naloxone enhanced it. Propranolol, diphenhydramine, cimetidine, atropine, *d*-amphetamine, fenfluramine and diazepam had little consistent effect on the satiety effect induced by CCK-8 (Wilson *et al.* 1983).

(1) *Conditioned aversion*

Peripheral administration of CCK may induce malaise, gastrointestinal dysfunction or produce aversive consequences, effects which might interfere with feeding responses. The hypothesis that CCK may induce satiety rests on evidence that the administration of exogenous CCK produced a normal behavioural sequence of satiety without evident ill-effects. The conditioned taste aversion (CTA) paradigm has been used to test the possibility that CCK administration has an aversive effect. If a flavour which has previously been paired with CCK administration is subsequently avoided, then it can be supposed that the CCK administration was aversive. Several studies testing this idea have failed to find evidence for a CTA following CCK-flavour pairings (Anika *et al.* 1981; Holt, Antin, Gibbs, Young, and Smith 1974; Houpt *et al.* 1978; Kraly *et al.* 1978; West, Williams, Braget, and Woods 1982).

Nevertheless, Deutsch and Hardy (1977) advanced evidence that CCK could act as an unconditioned stimulus in the formation of a CTA. Kraly *et al.* (1978), in addition, reported that partially pure CCK (20 per cent) produced a CTA under conditions in which CCK administration reduced food consumption. On the other hand, they could find no evidence that CCK-8 served as an unconditioned stimulus for a CTA.

The reinforcing or aversive consequences of cholecystokinin can also be estimated using a conditioned place preference paradigm. A recent experiment demonstrated that rats learned to avoid a distinctive environment which had previously been paired with CCK-8 (Swerdlow, van der Kooy, Koob, and Wenger 1983). The result points to aversive properties of CCK administration in the rat. The authors suggest that the decreased food intake following the administration of CCK in deprived animals is due to illness caused by the CCK. This suggestion denies CCK a physiological role in the normal process of satiety.

Several other pieces of evidence argue against a specific effect of peripherally-administered CCK to suppress feeding by generating a satiety signal. Thus in mice, CCK reduced exploratory activity (Crawley, Hays, Paul, and Goodwin 1981). In pigs, intravenous infusions of CCK-8 produced dose-related decrements in operant response rates when food, water, sucrose or heat were used as reinforcers (Baldwin *et al.* 1983). An earlier study by Mineka and Snowdon (1978) also called into question a physiological satiety function for CCK, because of inconsistent responses with repeated injections of CCK.

Arguments in favour of a relatively specific effect of CCK on feeding responses include the many observations that food consumption is reduced by CCK, when water intake is not. Furthermore, in mice, the ED_{50} for CCK's inhibitory effect on feeding is lower than those for a variety of other actions of the peptide (Zetler 1981). In human subjects who have been treated with CCK-8, the reports of sickness do not inevitably accompany the reductions in food consumption associated with the peptide treatment (Kissileff *et al.* 1981; Pi-Sunyer *et al.* 1982). It remains possible that any aversive consequence of CCK administration may be dissociable from an effect to reduce food consumption specifically.

(m) *Central CCK and feeding*

Much of the work concerned with the feeding suppressant effects of CCK has rested on peripheral administration of the peptide. Ideas about the mechanisms of action of CCK's satiety signal have emphasized release of endogenous CCK from the gut as a result of food ingestion. Peripheral circulating CCK is unlikely to cross the blood-brain barrier, and therefore its sites of action are peripheral to the brain too.

With the discovery of an extensive system of CCK-containing neurones in the central nervous system, it becomes of particular interest to establish the contribution, if any, of central CCK to feeding processes. The preliminary data now available suggest that central CCK may play some part in the expression of feeding activity.

Thus, intraventricular administration of CCK reduced food consumption in sheep (Della-Fera and Baile 1979, 1980), reduced operant responding for food reward in hungry pigs (Parrott and Baldwin 1981), and reduced food consumption in hens and chicks (Denbow and Myers 1982; Savory and Gentle 1983). In sheep, continuous infusion of the CCK receptor antagonist, dibutyryl cyclic GMP, into the cerebral ventricles produced a large increase in feeding (Della-Fera, Baile and Peikin 1981).

In the rat, however, intracerebroventricular administration of CCK has proven to be relatively ineffective in suppressing food intake (Della-Fera and Baile 1979; Lorenz and Goldman 1982; Nemeroff *et al.* 1978). In one study, intraventricular CCK reduced operant responding for food only after a period of 60 min had elapsed (Maddison 1977). Nevertheless, there is evidence that CCK is effective is suppressing noradrenaline-induced feeding when the peptide was infused directly into hypothalamic or preoptic area sites (McCaleb and Myers 1980; Myers and McCaleb 1981).

These data together suggest that actions of CCK at central sites can influence feeding responses. The data are only preliminary however, and it is not yet clear that there is a specific association between central CCK activity, in at least some brain regions, and the normal controls of feeding behaviour.

Other evidence encourages the search for a possible link. Thus, Straus and Yalow (1979, 1980) reported that levels of CCK were decreased in the cerebral cortex of fasted and of genetically obese mice. Against these reports, Sneider, Monahan, and Hirsch (1979) reported normal CCK concentrations in genetically obese rats and mice, in food deprived rats, and in obese animals with lesions in the ventromedial hypothalamus. Cholecystokinin receptor sites may vary in number as a function of deprivation state or obesity. Deprivation of food for 42 h significantly increased the numbers of CCK receptors in the olfactory bulb and hypothalamus of mice (Saito, Williams, and Goldfine 1981). There was a significant increase in the number of binding sites in cerebral cortex, hypothalamus, and olfactory lobes in mice made obese with goldthioglucose (Saito, Williams, Waxler, and Goldfine 1982). The significance of these changes is not known, but they add some weight to the possibility that central CCK-containing neurones and CCK receptor sites are implicated in feeding, hunger, and the control of body weight. Clearly, the concern which has been expressed regarding the specificity of CCK's peripheral actions in relation to feeding extends to central experiments. Considerably more detailed work is required before an involvement of central CCK in feeding-related processes can be adequately assessed.

Water intake

In contrast to the well-documented suppressant effects of CCK on food consumption, many investigations have failed to detect comparable effects of CCK on water consumption (Anika *et al.* 1977; Bernstein *et al.* 1976; Gibbs *et al.* 1973*a*; Kraly *et al.* 1978; Mueller and Hsiao 1977; Parrott and Baldwin 1981).

BOMBESIN

Isolation

Bombesin was first isolated by Vittorio Erspamer and his colleagues from the skin of two European frogs, *Bombina bombina* and *Bombina variegata variegata*, in the course of systematic investigations of amphibian tissue for biologically-active peptide families (Anastasi, Erspamer, and Bucci 1971; Erspamer 1980). In addition, several related peptides (amongst them, alytesin, litorin, and ranatensin) have been isolated from the skin of amphibians (Anastasi, Erspamer, and Endean 1975; Erspamer, Melchiorri, Falconieri, Erspamer and Negri 1978). The amino acid sequences of these peptides are shown in Table 2.4. McDonald and co-workers have isolated a mammalian bombesin-like peptide from porcine non-antral gastric tissue (McDonald, Jörnvall, Nilsson, Vagne, Ghatei, Bloom, and Mutt 1979). This heptacosapeptide is a potent releaser of gastrin, a property which it shares with bombesin. The amino acid sequence of porcine gastric gastrin-releasing peptide is also shown in Table 2.4. Nine of the ten *C*-terminal amino acids of porcine gastric gastrin-releasing peptide are identical with those of the corresponding *C*-terminal amino acid residues of amphibian bombesin.

Localization

Bombesin and bombesin-like peptides have been identified in tissue of fish, amphibians, birds, and mammals. In the rat gastrointestinal tract, highest concentrations of bombesin-like immunoreactivity have been found in the acid-secreting part of the stomach. Lower concentrations were detected in the

TABLE 2.4 *Amino-acid sequence of bombesin and related peptides*

Bombesin	*p*Glu-Gln-Arg-Leu-Gly-Asn-Gln-Trp-Ala-Val-Gly-His-Leu-Met-NH_2
Alytesin	*p*Glu-Gly-Arg-Leu-Gly-Thr-Gln-Trp-Ala-Val-Gly-His-Leu-Met-NH_2
Litorin	*p*Glu-Gln-Trp-Ala-Val-Gly-His-Phe-Met-NH_2
Ranatesin	*p*Glu-Val-Pro-Gln-Trp-Ala-Val-Gly-His-Phe-Met-NH_2
Gastrin-releasing peptide	Ala-Pro-Val-Ser-Val-Gly-Gly-Gly-Thr-Val-Leu-Ala-Lys-Met-Tyr-Pro-Arg-Gly-Asn-His-Trp-Ala-Val-Gly-His-Leu-Met-NH_2

antrum and jejunum (Dockray, Vaillant, and Walsh, 1979). Immunohistochemical study of the rat stomach showed that the bombesin-like reactivity was localized in nerve fibres in the myenteric plexus, submucosal plexus, and in the mucosa (Dockray *et al.* 1979). In the small and large intestine, bombesin-like immunoreactivity was observed principally in nerve fibres in the myenteric plexus. In contrast, in the gastrointestinal tract of frogs, fish and birds, bombesin appears to be localized in endocrine-like cells. Bombesin-like immunoreactivity has been identified in the human gastrointestinal tract (Polak, Bloom, Hobbs, Solcia, and Pearse 1976). Bombesin-like peptides have not been identified in rat liver, spleen, kidneys, pancreas, adrenal, pituitary, or pineal glands (Nemeroff, Luttinger, and Prange 1983).

Bombesin-like immunoreactivity has been detected in the brain of rats, dogs, and sheep (Moody and Pert 1979; Villarreal and Brown 1978; Walsh, Wong, and Dockray 1979). Using radioimmunoassay, Moody, Pert, and Jacobowitz (1979) undertook a comprehensive regional analysis of bombesin-like immunoreactivity in the rat central nervous system. Highest concentrations were detected in the substantia gelatinosa and nucleus tractus solitarius of the hindbrain. High concentrations were also observed in the interpeduncular nucleus of the midbrain and the arcuate nucleus of the hypothalamus. Limbic regions, including nucleus accumbens, septem, stria terminalis, and the central nucleus of the amygdala, contained intermediate concentrations. Lower concentrations were found in caudate nucleus, hippocampus, and cingulate cortex.

Binding sites

Using $[^{125}I]$-bombesin, high-affinity bombesin binding has been demonstrated in rat brain. The binding was specific, saturable and reversible (Moody, Pert, Rivier, and Brown 1978). Bombesin binding sites are heterogeneously distributed throughout the brain. In the hypothalamus, thalamus, midbrain, pons-medulla, and cerebellum, there is a strong correlation between the density of binding sites and the level of bombesin-like immunoreactivity. However, there are three regions where the positive correlation does not hold. In hippocampus, striatum and cortex, there are high densities of bombesin binding sites coupled with low concentrations of bombesin-like immunoreactivity.

Peripheral effects

Systemically-adminstered bombesin elicits a hypertensive response in many species (Erspamer, Melchiorri, and Sopranzi 1972). In addition, it increases muscle tone and rhythmic movements in several tissues, has an antidiuretic action and increases renin secretion (Erspamer 1980). Bombesin enhances the release of the pancreatic hormones, insulin, glucagon, and pancreatic polypeptide (Nemeroff *et al.* 1983). It stimulates gastrin release and gastric-acid secretion, and brings about increased plasma cholecystokinin levels (Erspamer 1980).

Central effects

Intracisternal administration of bombesin produces hyperglycemia in anaesthetized and unanaesthetized rats (Brown, Rivier, and Vale 1977*b*; Brown, Taché, and Fisher 1979). The available evidence indicates that bombesin may act centrally to increase sympathetic outflow which brings about the release of adrenaline from the adrenal medulla. The effects of this release are to lower plasma insulin levels and to raise plasma glucagon. The net outcome is a hyperglycaemia (Brown 1981).

Bombesin administered centrally also has marked effects on thermoregulation in rats and mice. Thus, it produces hypothermia in cold-exposed animals (Brown, Rivier, and Vale 1977*a*; Mason, Nemeroff, Luttinger, Hatley, and Prange 1980).

Intracerebroventricular injection of bombesin induces excessive grooming in mice (Katz 1980) and rats (de Caro, Massi, and Micossi 1980*e*; Kulkosky, Gibbs, and Smith 1982). It has also been reported to increase locomotor activity in rats (Pert, Moody, Pert, DeWald, and Rivier 1980) and to decrease rearing in an open-field (Cantalamessa, de Caro, Massi, and Micossi 1982).

Effects of bombesin on feeding and drinking need to be considered in relation to its manifold peripheral and central actions.

Food intake

Gibbs and his colleagues examined the effects of intraperitoneally administered bombesin in adult male rats which were given access to a liquid diet following a 3 h period of food deprivation. They discovered that bombesin (2–15 μg/kg) produced a dose-related suppression of the liquid food intake (Gibbs, Fauser, Rowe, Rolls, Rolls, and Maddison 1979). They observed a similiar effect of bombesin when solid food pellets were offered to 7-h food-deprived animals.

Several pieces of evidence led the authors to consider that bombesin exerts a satiety effect to reduce food consumption. Thus bombesin did not delay initiation of feeding, or slow the initial phase of eating. Animals stopped eating sooner following bombesin. Bombesin did not elicit any abnormal behaviour, or produce obvious signs of illness or distress. Furthermore, the largest dose of bombesin tested did not affect water consumption in 6-h water-deprived animals. This latter observation rules out a general behavioural suppression or debilitation as the explanation for bombesin's effect on food intake. The authors suggested that bombesin may serve as a satiety signal to bring about an early finish to feeding.

There have been many confirmatory-reports that exogenously-administered bombesin suppresses food consumption in lean rats. In addition, bombesin has been shown to attenuate food intake in obese animals, in weanling Zucker obese rats (McLaughlin and Baile 1980*a*), in weanling and adult obese (ob/ob) mice (McLaughlin and Baile 1981), and in obese rats with lesions of the ventromedial hypothalamus (West *et al.* 1982).

The mammalian peptide which resembles bombesin, gastrin-releasing peptide (Table 2.4) has also been investigated for its effects on food intake in rats. Stein and Woods (1982) reported that gastrin-releasing peptide produces a dose-dependent decrease in 30-min food intake of mildly food-deprived rats given access to a liquid diet. It was approximately 30 per cent less potent on a molar basis than bombesin in its suppression of food intake.

The mechanism(s) by which systemic administration of bombesin reduces food consumption remain unidentified. Martin and Gibbs (1980) demonstrated that intraperitoneal injections of bombesin retain effectiveness in reducing food consumption when tested in sham-feeding rats. Bombesin elicited the characteristic behavioural sequence of satiety in these animals. Like CCK, Martin and Gibbs (1980) proposed that exogenous bombesin is a sufficient stimulus for satiety in the sham-feeding animal, in which the contributions of intestinal and post-absorptive satiety mechanisms are excluded.

Unlike the case of CCK, however, the satiety effect of bombesin was not blocked by bilateral subdiaphragmatic vagotomy (Smith, Jerome, and Gibbs 1981). These results indicate that bombesin's satiety effect does not depend on an intact abdominal vagus. Other evidence suggests that bombesin and CCK may act through independent mechanisms to reduce food consumption. The joint effects of bombesin and CCK administered together are strictly additive (Gibbs, Kulkosky, and Smith 1981; Stein and Woods 1981). Differences also occur in the feeding responses of weanling Zucker obese rats to CCK and bombesin challenges (McLaughlin and Baile 1980*a*).

Adrenalectomy does not affect the satiety effect of bombesin in rats (Gibbs *et al.* 1981; Morley and Levine 1981). The hyperglycaemic response to CCK, mediated by the adrenal medulla, cannot account, therefore, for the reduction in food consumption which is produced by bombesin.

A recent study by Kraly, Miller, and Gibbs (1983) reveals a degree of complexity in the satiating effects of bomesin, which will have to be taken into account in interpreting the mechanisms of bombesin's actions on feeding responses. Following 24-h food deprivation, intraperitoneal administration of bombesin inhibited food intake in a dose-related manner during the day, whereas the inhibition of feeding observed during the night was not dose-related.

Following 3-h food deprivation, exogenous bombesin inhibited food intake in a dose-related manner at night, but not as clearly so during the day. Evidently the degree of hunger and the time of the day are important factors in determining the degree of satiety associated with bombesin's actions.

It remains possible that exogenously-administered bombesin produces malaise, or is aversive in some other sense, and thereby reduces feeding. Deutsch and Parsons (1981), in fact, provided evidence that bombesin (16 μg/kg) injected intraperitoneally induced a conditioned taste aversion in a two-bottle test. They argued that bombesin reduces food intake, not through satiation, but as a consequence of its aversiveness when administered systemically. In contrast, there are two additional studies which failed to detect a conditioned taste aversion with bombesin. In one of the studies, 4 μg/kg bombesin was administered intraperitoneally (Kulkosky, Gray, Gibbs, and Smith 1981), and in the other, 4 μg/rat in rats weighing on average less than 250 g (West *et al.* 1982). There are procedural differences amongst the three studies, and it is difficult to determine the basis for the discrepancy separating the first from the latter two reports. The issue requires further enquiry to establish what conditions are necessary for a bombesin-induced conditioned taste aversion to be obtained.

Central administration of bombesin has been shown to reduce food consumption in rats (de Caro *et al.* 1980*e*; Morley and Levine 1980; Stuckey and Gibbs 1981; Woods, West, Stein, McKay, Lotter, Porte, Kenney, and Porte 1981), and in sheep (Baile and Della-Fera 1981). Nevertheless, there is convincing evidence that intracerebroventricular administration of bombesin affects feeding responses as a consequence of the induction of abnormal behaviour. De Caro *et al.* (1980*e*) reported that 1, 5, and 10 μg/rat bombesin produced a dose-related suppression of food intake in hungry rats. However, the effect on feeding coincided with the induction of excessive grooming which would be expected to interfere with the ingestive response. These observations have been confirmed (Cantalamessa *et al.* 1982; Kulkosky *et al.* 1982). The effects of intracerebroventricular injection of bombesin on food intake cannot, therefore, be taken as evidence for a feeding-satiety action of the peptide. The preliminary report that bombesin applied to discrete lateral hypothalamic loci can reduce feeding without greatly increasing grooming (Stuckey and Gibbs 1981) suggests that the effects of bombesin on feeding are separable from the induction of grooming. Such evidence is critical if centrally-acting bombesin is to be accorded a specific role in the satiation of feeding responses.

Water intake

Surprisingly, intracerebroventricular administration of bombesin elicits reliable and copious drinking in the pigeon (de Caro, Massi, and Micossi 1980*b*) and in the Peking duck, *Anas plathyrinos plathyrinos* (de Caro, Mariotti, Massi, and Micossi 1980*c*). In the pigeon, bombesin (6–600 pmol per bird) produced dose-related drinking, but did not induce eating and did not affect the cloacal temperature of the birds. Dose-dependent effects of bombesin were observed in the duck (6–3000 pmol per bird). In both avian species, comparisons of the potencies of several dipsogenic peptides yielded the following rank order: angiotensin II > bombesin > substance P. In a study of structure–activity relationships, it was found that elimination of six or more *N*-terminal amino acids from the bombesin molecule dramatically reduced the dipsogenic effect of the peptide (de Caro, Mariotti, Massi, and Micossi 1980*d*).

Angiotensin II analogues, which antagonize the dipsogenic effects of intracerebroventricular infusion of angiotensin II in the pigeon, proved ineffective against the drinking produced by bombesin (de Caro, Massi, Micossi, and Perfumi 1982). These results indicate that the drinking response elicited by bombesin in pigeons is mediated by mechanism(s) which do not involve the activation of angiotensin II receptors. The mechanism(s) which subserve bombesin-induced drinking remain to be identified.

In contrast to these findings in birds, intracerebroventricular administration of bombesin in the rat suppresses drinking. Thus, bombesin has been shown to inhibit drinking induced by 16-h water deprivation, central administration of angiotensin II (100 ng), or central administration of carbachol (300 ng) (de Caro *et al.* 1980*e*). Bombesin (7.8 ng/rat) inhibited angiotensin-induced drinking by 22.9 per cent; at 500 ng/rat, the drinking was inhibited by 88.2 per cent. The authors also noted that food intake in hungry rats was inhibited by bombesin in doses of 1–10 μg/rat, but also pointed out that in this dose range, bombesin induced excessive grooming. Cantalamessa *et al.* (1982) examined the behavioural specificity of the antidipsogenic effect of bombesin in greater detail. They showed that in larger doses, intracerebroventricular bombesin suppressed feeding and drinking, induced excessive grooming, and affected behaviour in an open field. Nevertheless, 7.81 ng/rat bombesin was sufficient to inhibit the drinking elicited by angiotensin II to a significant extent. At this dose, bombesin did not produce other behavioural alterations. Hence, the authors concluded that bombesin can exert a specific antidipsogenic effect in the rat.

GLUCAGON

Glucagon is familiar as a pancreatic hormone with a glycogenolytic–hyperglycaemic effect. It is also found in the gut (Moody, Jacobsen, and Sundby, 1978).

It has been proposed that pancreatic glucagon may act as a satiety hormone involved in the production of the satiety which follows a meal. Endogenous glucagon levels appear to increase during meals (De Jong, Strubbe, and Steffens 1977; Goriya, Bahoric, Marliss, Zinman, and Albisser 1981; Unger and Orci 1976). In humans, the administration of exogenous glucagon reduces feeding (Penick and Hinkle 1961; Penick and Smith 1964; Schulman, Carleton, Whitney, and Whitehorn 1957). Similarly, when glucagon was administered just before meals, rats consumed less food by finishing the meal sooner; the initial rate of feeding was unimpaired (Geary and Smith 1982*a*). The feeding suppressant effect of glucagon was specific, since it did not affect water intake or body temperature, and did not produce a conditioned taste aversion (Geary and Smith 1982*a*; Martin and Novin 1977).

Both cholecystokinin (Gibbs, Young, and Smith 1973*b*) and bombesin (Martin and Gibbs 1980) elicit feeding satiety in sham-feeding rats. Interestingly, glucagon failed to suppress consumption in the sham-feeding rat (Geary and Smith 1982*b*). Glucagon elicited marked hyperglycaemia during sham-feeding, and therefore the hyperglycaemia which normally accompanies ingesting a meal may not be critical to glucagon's satiety effect.

The satiety effect of glucagon has been blocked by subdiaphragmatic vagotomy in rats and rabbits (Martin, Novin, and VanderWeele 1978; VanderWeele, Geiselman, and Novin 1979). Selective disconnection of the hepatic branch of the abdominal vagus was sufficient to block glucagon's satiety effect (Geary and Smith 1983). Abdominal vagotomy which spared the hepatic branch, however, did not alter its effect. Since glucagon inhibited feeding after blockade of peripheral postganglionic muscarinic receptors with atropine, it is possible that hepatic vagal afferents are required for glucagon's satiety effect (Geary and Smith 1983). Again there is a contrast with CCK. Selective vagotomies indicate that the gastric branch of the vagus is necessary to the feeding suppressant effect of CCK.

NEUROTENSIN

Neurotensin was isolated from bovine hypothalamus (Carraway and Leeman 1975). When injected intravenously in the rat, it produced dose-dependent hypotension, vasodilation, and cyanosis. The amino-acid sequence of neurotensin is:-

pGlu–Leu–Tyr–Glu–Asn–Lys–Pro–Arg–Arg–Pro–Tyr–Ile–Leu–OH.

Xenopsin is an octapeptide that is structurally related to neurotensin. It was isolated from the skin of the frog *Xenopus laevis* (Araki, Tachibana, Uchiyama, Nakajima and Yasuhara 1973).

Neurotensin is found in very high concentration in tissues of the gastro-intestinal tract (Buchan, Polak, Sullivan, Bloom, Brown, and Pearse 1978), and has been detected in the pancreas (Berelowitz, Nakawatase, and Frohman 1980). Neurotensin exerts many effects in the periphery, including the production of hypotension, effects on smooth muscle contraction in the ileum, effects on cardiac activity, the production of hyperglycaemia, and effects on gastro-entero-pancreatic secretions (Nemeroff *et al.* 1983).

Neurotensin-like immunoreactivity is widely distributed in the brain. Highest concentrations in the rat are found in the nucleus accumbens, preoptic area, ventromedial and periventricular hypothalamic nuclei (Kobayashi, Brown, and Vale 1977). Using an autoradiographic technique, Young and Kuhar (1979, 1981) have described high densities of neurotensin receptors in the olfactory bulb, cingulate cortex, substantia nigra, septum, amygdaloid nuclei, nucleus accumbens and several additional regions.

Neurotensin also exerts several central effects. Intracisternal administration of neurotensin produced a marked hypothermia in cold-exposed mice (Bissette,

Nemeroff, Loosen, Prange, and Lipton 1976). Clineschmidt and McGuffin (1977) reported that low doses of neurotensin administered intracisternally had an antinociceptive effect in female mice. Centrally-administered neurotensin has also been reported to produce hypotension in rats (Rioux, Quirion, St. Pierre, Regoli, Jolicoeur, Belanger, and Barbeau 1981).

Neurotensin suppresses food consumption in rats. Thus intracerebroventricular administration of neurotensin (3.3-30 μg) has been reported to produce a dose-related reduction in food consumption in food deprived rats (Luttinger, King, Sheppard, Strupp, Nemeroff, and Prange 1982). Bilateral injections of neurotensin into the paraventricular nucleus of the hypothalamus have been shown to suppress feeding in food deprived rats, but not drinking in thirsty animals (Hoebel, Hernandez, McLean, Stanley, Aulissi, Glimcher, and Margolin 1982). Intracerebroventricular neurotensin (30 μg) did not produce a conditioned taste aversion (Luttinger *et al.* 1982).

In a series of experiments, Levine, Kneip, Grace, and Morley (1983) confirmed that intraventricular injection of neurotensin suppressed food consumption in food deprived rats. They showed that it also reduced spontaneous feeding in vagotomized rats, indicating that an intact vagus in not required for neurotensin's suppressant effect on feeding. Neurotensin suppressed the feeding induced by intracerebroventricular injections of noradrenaline or dynorphin, but not that induced by intracerebroventricular muscimol or subcutaneously administered insulin. These authors discuss the points of similarity between the effects of neurotensin and bombesin on feeding responses.

SOMATOSTATIN

The presence of a substance in hypothalamic extracts that inhibits the basal release of pituitary growth hormone from rats was first shown by Krulich, Dhariwal and McCann (1968) and Krulich and McCann (1969). Subsequently, a peptide was isolated from pig and sheep hypothalami that was shown actively to inhibit GH release *in vitro* and *in vivo* (Brazeau, Vale, Burgus, Ling, Butcher, Rivier, and Guillemin 1973; Schally, Dupont, Arimura, Redding, Nishi, Linthicum, and Schlesinger 1976). Somatostatin is a cyclic tetradecapeptide, and its amino acid sequence is:-

H-Ala-Gly-Cys-Lys-Asn-Phe-Phe-Trp-Lys-Thr-Phe-Thr-Ser-Cys-OH
(disulfide bridge linking Cys 3 and Cys 14 shown by bracket)

Immunoreactive somatostatin has been demonstrated in the stomach, intestines, salivary glands and pancreas (Arimura, Sato, Dupont, Nishi, and Schally 1975; Hökfelt, Efendic, Hellerstrom, Johansson, Luft, and Arimura 1975). Somatostatin suppresses the release of insulin and glucagon from the pancreas, and inhibits the release of most gastrointestinal hormones, including gastrin, secretin, cholecystokinin, motilin and pancreatic polypeptide (Konturek 1980).

In addition to inhibiting the release of pituitary GH, somatostatin also inhibits the release of pituitary thyroid stimulating hormone (TSH), which is induced by thyrotropin releasing hormone (TRH). Somatostatin does not, however, affect basal release of prolactin, luteinizing hormone (LH), or adrenocorticotrophic hormone (ACTH) (Rorstad, Martin, and Terry 1980).

Somatostatin is widely and heterogeneously distributed in the central nervous system. Besides its presence in hypothalamic nuclei, somatostatin is also found in the cerebral cortex, hippocampus, basal ganglia, and the limbic system (Palkovits and Brownstein 1983).

Somatostatin has been shown to decrease food intake, but not water consumption in rats and baboons. In rats deprived of food for 5-6 h in the light phase, somatostatin (10 ng/kg-1 μg/kg, administered intraperitoneally) reliably reduced food consumption (Lotter, Krinsky, McKay, Treneer, Porte, and Woods 1981). Somatostatin did not produce a conditioned taste aversion and did not reduce water intake in thirsty rats. Intracerebroventricular injection of somatostatin (100 ng/rat) did not affect feeding. Similiar findings were obtained in male baboons (Lotter *et al.* 1981). Intraperitoneally-administered somatostatin (1 μg/kg) decreased food consumption, although intracerebroventricular somatostatin (2 μg/baboon) had no effect. The authors suggest, on the basis of these findings, that somatostatin may act peripherally as a satiety factor.

Levine and Morley (1982) confirmed that the intraperitoneal administration of somatostatin (10 μg/kg) reduced spontaneous nocturnal feeding in rats. This effect was abolished in vagotomized animals, which suggests that its suppressant effect on feeding may be mediated by the vagus. Somatostatin also inhibited stress (tail-pinch)-induced eating, and insulin-induced feeding. It failed, however, to affect the feeding induced by intracerebroventricular administration of noradrenaline, muscimol, or dynorphin. It also failed to reduce food intake in food-deprived animals. Levine and Morley (1982) and Lotter *et al.* (1981) drew attention to the many parallels between the effects of somatostatin and CCK.

THYROTROPIN RELEASING HORMONE

Thyrotropin releasing hormone (TRH) is located in hypothalamic neurosecretory cells, and was the first example of the hypothalamic releasing hormones to be isolated and characterized. Its name derives from its role in the secretion of thyroid stimulating hormone (TSH) from the anterior pituitary. TRH is a tripeptide, and its structure is L-pyroglutamyl-L-histidyl-L-prolineamide (*p*Glu-His-Pro·NH_2).

The functions of TRH are not, however, limited to the control of TSH release. TRH is widely distributed through the extrahypothalamic nervous system and spinal cord, and is also found in the retina, pancreas, gastrointestinal tract, placenta and adrenals (Morley 1979). TRH exerts a variety of effects, both centrally and peripherally.

Vijayan and McCann (1977) reported that TRH injected into the third ventricle of food and water-deprived rats suppressed food and water intake. Drinking was completely suppressed when TRH was given in a dose of 1.5 nmoles.

Intraperitoneal and subcutaneous administration of TRH has also been reported to suppress food intake in food-deprived rats (Morley and Levine 1980; Vogel, Cooper, Barlow, Prange, Mueller, and Breese 1979).

Morley, Levine and their colleagues examined the effects of intracerebroventricular injection of TRH in a range of feeding paradigms. They found that TRH inhibited stress (tail-pinch)-induced feeding (Morley and Levine 1980), but failed to affect the feeding induced by intraventricular administration of muscimol (Morley *et al.* 1981), noradrenaline (Morley *et al.* 1982*c*), dynorphin(1-13) or bromergocryptine (Morley *et al.* 1982*a*).

In vagotomized animals, TRH injected intraperitoneally failed to reduce food intake induced by food deprivation (Morley *et al.* 1982*b*). However, centrally administered TRH significantly reduced food intake. In sham-feeding animals, TRH reduced feeding by either route. The authors suggest that TRH may have two sites of action, one central and the other peripheral, in its suppressant effect of food ingestion. Peripherally-injected TRH produced an effect on feeding which depended on an intact vagus.

CORTICOTROPIN-RELEASING FACTOR

Corticotropin-releasing factor (CRF) is a 41 amino-acid residue peptide which has been extracted from sheep hypothalamus (Spiess, Rivier, Rivier, and Vale 1981; Vale, Spiess, Rivier, and Rivier 1981). CRF potently stimulates the release of adrenocorticotrophic hormone (ACTH) and β-endorphin from the pituitary *in vivo* and *in vitro* (Rivier, Brownstein, Spiess, Rivier, and Vale 1982). CRF has been confirmed in the hypothalamus, and may also be present in extrahypothalamic locations (Palkovits and Brownstein 1983). Central administration of CRF produced a variety of effects, including increased plasma levels of catecholamines, glucose and vasopressin. Arterial pressure, heart rate and oxygen consumption were elevated (Brown, Fisher, Rivier, Spiess, Rivier, and Vale 1982).

Morley and Levine (1982) found that intraventricular administration of CRF decreased food consumption during the night, and also reduced the feeding response aroused by food deprivation. CRF administration also reduced fluid consumption in thirsty rats. These investigators also showed that intraventricular CRF suppressed feeding which was induced by muscimol, noradrenaline, dynorphin and insulin (Levine, Rogers, Kneip, Grace, and Morley 1983). However, CRF produced a marked increase in grooming, and it is quite possible that the reductions in feeding and drinking occurred as a result of the excessive grooming. In this respect, the effects of intraventricular injections of CRF resemble those of bombesin, which in large doses induces excessive grooming, at the expense of feeding and drinking.

Sauvagine is a 40-amino-acid-residue polypeptide which has been recently isolated from the skin of the frog *Phyllomedusa sauvagei* (Erspamer, Falconieri, Erspamer, Improta, Negri, and de Castiglione 1980; Montecucchi and Henschen 1981). Like CRF, sauvagine stimulates the release of ACTH and β-endorphin (Brown, Fisher, Spiess, Rivier, Rivier, and Vale 1982). In a comparison between the effects of CRF and sauvagine on feeding responses in rats, Gosnell, Morley, and Levine (1983) have shown that both peptides administered intracerebroventricularly reduced spontaneous nocturnal feeding, feeding induced by 24–25-h food deprivation, and feeding induced by the kappa-opiate receptor agonist, ethylketocyclazone. Although these data may indicate satiety effects of both peptides, the authors also establish that they were effective as unconditioned stimuli in producing conditioned taste aversion. Consequently, there may have been aversive effects of the injections which could have affected the feeding responses.

SUBSTANCE P AND OTHER TACHYKININS

Isolation

Tachykinins constitute a family of peptides, some of which were first identified in amphibian skin. The first of the group to be isolated was physalaemin, obtained from the skin of the South American frog, *Physalaemus bigilonigerus.* Others include uperolein from the frog, *Uperoleia rugosa,* phyllomedusin from the Amazonian frog *Phyllomedusa bicolor,* and kassinin from African frogs (*Kassina senegalensis*) (Bertaccini 1980; Erspamer 1981).

The related peptide eledoisin was obtained from salivary glands of two Mediterranean octopods *Eledone moschata* and *Eledone aldrovandi.*

Von Euler and Gaddum (1931) first described a substance extracted from mammalian brain and intestine which had a vasodepressor action. Forty years later, substance P was isolated (Chang and Leeman 1970), and its amino acid sequence determined (Chang, Leeman, and Naill 1971). Table 2.5 indicates the structures of this family group of peptides.

Substance P distribution

The distribution of substance P was first studied by bioassay (Amin, Crawford, and Gaddum, 1954; Lembeck, 1953; Pernow, 1953). Substance P was described in the spinal dorsal roots, and Lembeck (1953) proposed that substance P might be the transmitter of primary sensory neurones. Recently, radioimmunoassay

TABLE 2.5 *Amino-acid sequences of substance P and related tachykinins*

Sequence	Peptide
Pyr-Ala-Asp-Pro-Asn-Lys-Phe-Tyr-Gly-Leu-Met-NH_2	Physalaemin
Pry-Pro-Asp-Pro-Asn-Ala-Phe-Tyr-Gly-Leu-Met-NH_2	Uperolein
Pyr-Pyr-Asn-Pro-Asn-Arg-Phe-Ile-Gly-Leu-Met-NH_2	Phyllomedusin
Asp-Val-Pro-Lys-Ser-Asp-Glu-Phe-Val-Gly-Leu-Met-NH_2	Kassinin
Pyr-Pro-Ser-Lys-Asp-Ala-Phe-Ile-Gly-Leu-Met-NH_2	Eledoisin
Arg-Pro-Lys-Pro-Gln-Gln-Phe-Phe-Gly-Leu-Met-NH_2	Substance P

and immunocytochemical methods have been employed to describe the localization and distribution of substance P in the central nervous system and elsewhere (Jessell 1983).

Immunocytochemical studies have shown substance P reactivity to be present in sympathetic ganglia (Hökfelt, Elfvin, Schultzberg, Goldstein, and Nilsson 1977), the adrenal medulla (Saria, Wilson, Molnar, Viveros, and Lembeck 1980), and in enteric neurones (Costa, Furness, Franco, Llewellyn-Smith, Murphy, and Beardsley 1982). Substance P is also located in visceral sensory neurones (Katz and Karten 1980), and in spinal and trigeminal sensory neurones (Cuello, del Fiacco, and Paxinos 1978; Hökfelt, Kellerth, Nilsson, and Pernow 1975).

Substance P is present in many neuronal pathways within the central nervous system (Iversen 1983; Jessell 1983). Major projections that have been investigated include a projection from the corpus striatum to the globus pallidus, interpeduncular nucleus, and substantia nigra, and a habenular-interpedunclar projection. Immunocytochemical studies have further revealed a great number of other substance P immunoreactive cell bodies, fibres and terminals (Ljungdahl, Hökfelt, and Nilsson 1978; Cuello and Kanazawa 1978).

Receptors

High affinity binding of [^{3}H]-substance P to rat brain membranes *in vitro* has been described (Hanley, Sandberg, Lee, Iversen, Brundish, and Wade 1980).

Substance P exerts a wide variety of pharmacological effects, and in each case it appears that full activity resides in the *C*-terminal hexapeptide sequence. The actions of substance P are mimicked by other tachykinins, which share the *C*-terminal sequence, – Phe–[ ]-Gly–Leu–Met–NH_2 (Table 2.5). Interestingly, when relative potencies amongst tachykinins are compared, there are two distinct patterns of tissue response. In the 'substance P-P' type, substance P, physalaemin, kassinin, and eledoisin are approximately equipotent; in the 'substance P-E' type, eledoisin and kassinin are up to several hundred times more potent than substance P and physalaemin (Erspamer 1981; Lee, Iversen, Hanley, and Sandberg 1982).

Central effects

High densities of substance P-containing terminals lie close to dopamine-containing neurones in the substantia nigra and interpeduncular nucleus. It has been proposed that substance P activates dopamine neurones which ascend to innervate the limbic system and basal ganglia, and so elicits behavioural effects which are known to be characteristic of dopamine-pathway activation (Iversen 1982).

Infusion of substance P into the ventral tegmental area causes increased locomotion in rats tested in an open field apparatus (Stinus, Kelly, and Iversen 1978). Selective destruction of the mesolimbic dopamine innervation using the neurotoxin, 6-hydroxydopamine (6-OHDA), or administration of haloperidol, a dopamine antagonist, blocked the effect of substance P (Kelley, Stinus, and Iversen 1979). These results prompt the suggestion that the effects of substance

P in the ventral tegmentum depend on an intact mesolimbic dopamine system. Experiments with a metabolically stable analogue of substance P, DiMe-C7 (pGlu^5,MePhe^8,Sar^9)-substance P (5–11), confirmed a locomotor stimulant effect when injected into the ventral tegmental area (Eison, Eison, and Iversen 1982).

Water intake

Angiotensin II, which is synthesized peripherally and in the central nervous system, acts as a potent dipsogen to elicit drinking when injected intravenously or intracerebrally. Drinking responses have been elicited by angiotensin II in all reptiles, birds and mammals that have been tested. Substance P and related tachykinins are particularly interesting, because following intracerebroventricular administration, they exert powerful antidipsogenic effects in rats, but act to promote drinking in pigeons and ducks (Evered 1983). This dissociation between the responses of birds and rats in relation to drinking behaviour is similiarly observed in the case of bombesin.

In the pigeon, intracranial injections of eledoisin, physalaemin, and substance P elicited drinking responses (de Caro, Massi, and Micossi 1978*c*; Evered, Fitzsimons, and de Caro 1977; Fitzsimons and Evered 1978). The dipsogenic potency of angiotensin II remained superior, with the potencies of physalaemin and eledoisin being similiar, and both markedly superior in effect to substance P. This pattern in the potency relationships does not match the distinctions, described above, between type-E and type-P responses to substance P. There may, of course, be another receptor type which mediates drinking responses in the pigeon, for which the potency relationships would be eledoisin = physalaemin > substance P.

Intracranial injections of vasopressin or bradykinin were ineffective in the pigeon (Evered *et al.* 1977).

The dipsogenic effects of the tachykinins have been clearly dissociated from their hypotensive effects, in studies which have examined potency relationships (de Caro, Massi, and Micossi 1980*a*), and which have investigated structure-activity relationships (de Caro, Massi, and Micossi 1979). Angiotensin II antagonists failed to affect the dipsogenic effect of eledoisin in pigeons, indicating that its effect on drinking behaviour was not mediated by angiotensin receptors (de Caro *et al.* 1982).

In ducks, intracerebroventricular injections of angiotensin II, eledoisin and bombesin each elicited a clear drinking response (de Caro, Mariotti, Massi, and Micossi 1980*c*). Substance P was ineffective in these animals.

In complete contrast to the findings in avian species, eledoisin, physalaemin and substance P are antidipsogenic in the rat when administered intraventricularly (de Caro, Massi, and Micossi 1978*a, b*; de Caro, Massi, Micossi, and Venturi, 1978*d*; Fitzsimons and Evered 1978; Morley, Levine, and Murray 1981).

Substance P had been shown to inhibit the drinking elicted by angiotensin II, carbachol, water deprivation and hypertonic sodium chloride (de Caro *et al.* 1978*a*). It was most effective in antagonizing the dipsogenic effect of angiotensin II. Physalaemin proved effective in inhibiting the drinking produced by angiotensin II, carbachol and water deprivation. Unlike substance P, it was ineffective in rats with a sodium chloride load (de Caro *et al.* 1978*d*). In rats challenged with angiotensin II as the dipsogenic stimulus, eledoisin was substantially more effective than physalaemin in its inhibition of drinking (de Caro *et al.* 1978*b*). In rats drinking in response to water deprivation, or in response to angiotensin II, potency relationships of eledoisin >> physalaemin > substance P have been established (Cantalamessa, de Caro, and Perfumi 1981; de Caro *et al.* 1980*a*). These potency relationships appear to indicate that a type-E, and not a type-P, response may underlie the antidipsogenic actions of tachykinins in rats.

In rats and pigeons, intraventricular administration of tachykinins produced marked decreases in arterial blood pressure (de Caro *et al.* 1980*a*). For rats, potency relationships for the hypotensive effect were physalaemin > eledoisin = substance P, whilst in the pigeon, they were substance P > physalaemin > eledoisin. The authors concluded that there could be no direct relationship between the effects of tachykinins on drinking responses and arterial blood pressure. Tachykinins induce vasopressin release, but this action may also be separated from their antidipsogenic effect in rats (Cantalamessa *et al.* 1981).

Food intake

Intraventricular administration of tachykinins does not affect feeding responses in rats adapted to a food deprivation schedule (de Caro *et al.* 1978*a,d*). Hence, the antidipsogenic effect of the tachykinins in this species is not due to general behavioural depression.

POSTSCRIPT

This chapter has considered evidence that CCK, bombesin, glucagon, neurotensin, somatostatin, TRH, CRF, and other neuropeptides may act as satiety signals and reduce food consumption. In addition, bombesin, substance P, and related tachykinins were considered in the light of their potent dipsogenic effects in birds, which contrast sharply with their antidipsogenic effects in rats.

It is becoming clear that neuropeptides which reduce food consumption do not share a single mechanism of action. For example, the suppressant effect of bombesin remained intact following vagotomy, while the effects of CCK, glucagon and somatostatin were abolished. Furthermore the feeding-suppressant effect of glucagon was shown to require the integrity of the hepatic branch of the vagus, while the effect of CCK depended on the retention of the gastric branch. A second example to illustrate the point draws on the results obtained in sham-feeding rats. Both CCK and bombesin remained effective in reducing food consumption in sham-feeding animals, while the effect of glucagon disappeared. As a third example, we can consider the effects of neuropeptides when administered intracerebroventricularly. Central administration of neurotensin, bombesin, TRH and CRF reduced feeding responses, but somatostatin and CCK (in the rat) had little or no effect on feeding. Comparisons between neuropeptides in this way are a necessary first step in the identification of possible differences underlying their mechanisms of action. Discrepancies between the effects of neuropeptides will provide leads to further detailed studies into the means by which these substances affect feeding responses.

The specificity of feeding suppressant effects requires careful evaluation for each putative satiety signal. In the cases of CCK and bombesin, for example, possible aversive consequences following their administration have been claimed. There is no reason *a priori* why administration of a peptide should not be aversive, as determined under a certain set of conditions, as well as acting as a satiety signal, as determined under a different set. An aversive effect, if found, may not be the explanation for a feeding suppressant effect. However, proponents of the idea that neuropeptides can function as specific satiety signals, and that their actions affecting feeding have physiological relevance, need to provide evidence that putative satiety and aversive effects are clearly dissociable.

Angiotensin II has been shown to be a particularly potent dipsogen in mammalian and bird species. It is puzzling that bombesin, substance P and related tachykinins, when administered intraventricularly, should similarly stimulate drinking in birds, but suppress drinking in the rat. Further comparisons would be interesting to determine if there is a general divergence between mammals and birds. We do not yet know how representative are the rat, pigeon, and duck (the animals investigated so far) of mammalian and avian species in general. Investigations into the mechanisms of action of these neuropeptides are required in order to judge the relevance of their observed effects on drinking to the physiological control of water and salt appetites, and of body fluid balance.

Neuropeptides are located in the central nervous system, in peripheral neurones and in many tissues including those of the gastrointestinal tract, skin, pituitary, pancreas and adrenals. As satiety signals, they have been thought of, primarily, acting as gut hormones released in response to food ingestion. This view of their actions may require radical revision. We are still too ignorant of the functions of neuropeptide function within the central nervous system to know how important it will be to consider central neuropeptide actions. Nevertheless, the possibility lies before us that neuropeptide actions at multiple locations, both within the central nervous system and without, will be relevant to many, if not all, aspects of the control and modulation of feeding and drinking responses.

Acknowledgements

I wish to thank Professors de Caro, Myers and Woods, Drs Baile, Evered, Levine, and Morley for providing me with reprints of their published work. Ms S. Garvey and Ms B. Humphries typed the manuscript.

References

Ajmone-Marsan, C. and Traczyk, W. Z. (eds.), (1980). *Neuropeptides and neural transmission*. International Brain Research Organization, Monograph series, Vol. 7. Raven Press, New York.

Akil, H. and Watson, S. J. (1983). Beta-endorphin and biosynthetically related peptides in the central nervous system. In *Neuropeptides* (eds. L. L. Iversen, S. D. Iversen, and S. H. Snyder) pp. 209–53. Plenum Press, New York.

Amin, A. H., Crawford, T. B. B., and Gaddum, J. H. (1954). The distribution of substance P and 5-hydroxytryptamine in the central nervous system of the dog. *J. Physiol. (Lond.)*. **126**, 596–618.

Anastasi, A., Erspamer, V., and Bucci, M. (1971). Isolation and structure of bombesin and alytesin, two analogous active peptides from the skin of the European amphibians, *Bombia* and *Alytes. Experientia* **27**, 166–7.

—— —— and Endean, R. (1975). Amino acid composition and sequence of litorin, a bombesin-like nonapeptide from the skin of the Australian Leptodactylid frog Litoria aurea. *Experientia* **31**, 510–11.

Anika, S. M., Houpt, T. R., and Houpt, K. A. (1977). Satiety elicited by cholecystokinin in intact and vagotomized rats. *Physiol. Behav.* **19**, 761–6.

—— —— —— (1981). Cholecystokinin and satiety in pigs. *Am. J. Physiol.* **240**, R310–R318.

Antin, J., Gibbs, J., Holt, J., Young, R., and Smith, G. P. (1975). Cholecystokinin elicits the complete behavioural sequence of satiety in rats. *J. Comp. Physiol. Psychol.* **89**, 784–90.

—— —— and Smith, G. P. (1977). Intestinal satiety requires pregastric food stimulation. *Physiol. Behav.* **18**, 421–5.

—— —— —— (1978). Cholecystokinin interacts with pregastric food stimulation to elicit satiety in the rat. *Physiol. Behav.* **20**, 67–70.

Araki, K., Tachibana, S., Uchiyama, M., Nakajima, T., and Yasuhara, T. (1973). Isolation and structure of a new active peptide 'xenopsin' on the smooth muscle, especially on a strip of fundus from a rat stomach, from the skin of *Xenopus laevis. Chem. Pharm. Bull.* **21**, 2801–4.

Arimura, A., Sato, H., Dupont, A., Nishi, N., and Schally, A. V. (1975). Somatostatin: abundance of immunoreactive hormone in rat stomach and pancreas. *Science* **189**, 1007–9.

Baile, C. A. and Della-Fera, M. A. (1981). Bombesin injected into cerebral ventricles decreases feed intake of sheep. *Fedn. Proc.* **40**, 308.

Baldwin, B. A., Cooper, T. R., and Parrott, R. F. (1983). Intravenous cholecystokinin octapeptide in pigs reduces operant responding for food, water, sucrose solution or radiant heat. *Physiol. Behav.* **30**, 399–403.

Barker, J. L. and Smith, T. G. (eds.). (1980). *The role of peptides in neuronal function*. Marcel Dekker, New York.

Batt, R. A. L. (1983). Decreased food intake in response to cholecystokinin (pancreozymin) in wild-type and obese mice (genotype ob/ob). *Int. J. Obesity*. **7**, 25–9.

Becker, E. E. and Grinker, J. A. (1977). Meal patterns in the genetically obese Zucker rat. *Physiol. Behav.* **18**, 685–92.

Beinfeld, M. C. (1983). Cholecystokinin in the central nervous system: a mini-review. *Neuropeptides* **3**, 411–27.

Berelowitz, M., Nakawatase, C., and Frohman, L. A. (1980). Pancreatic immunoreactive neurotensin (IR-NT) in obese (ob) and diabetes (db) mutant mice: a longitudinal study. *Diabetes* **29**, 55A.

Bernstein, I. L., Lotter, E. C., and Zimmerman, J. C. (1976). Cholecystokinin-induced satiety in weanling rats. *Physiol. Behav.* **17**, 541–3.

Bertaccini, G. (1980). Peptides of the amphibian skin active on the gut. I. Tachykinins (substance P-like peptides) and ceruleins. Isolation, structure and basic functions. In *Gastrointestinal hormones* (ed. G. B. J. Glass) pp. 315-341. Raven Press, New York.

Bissette, G., Nemeroff, C. B., Loosen, P. J., Prange, A. J., and Lipton, M. A. (1976). Hypothermia and intolerance to cold induced by intracisternal administration of the hypothalamic peptide neurotensin. *Nature* **262**, 607–9.

Bloom, F. E. (1983). The endorphins: a growing family of pharmacologically pertinent peptides. *Ann. Rev. Pharmacol. Toxicol.* **23**, 151–70.

Blundell, J. E. (1981). Bio-grammar of feeding: pharmacological manipulations and their interpretations. In *Theory in psychopharmacology* (ed. S. J. Cooper) Vol. 1, pp. 233–76. Academic Press, London.

—— and Leshem, M. B. (1975). Analysis of the mode of action of anorexic drugs. In *Recent advances in obesity research. I.* (ed. A. Howard) pp. 368–71. Newman, London.

Bolles, R. C. (1960). Grooming behaviour in the rat. *J. Comp. Physiol. Psychol.* **53**, 306–10.

Brazeau, P., Vale, W., Burgus, R., Ling, N., Butcher, M., Rivier, J., and Guillemin, R. (1973). Hypothalamic peptide that inhibits the secretion of immunoreactive pituitary growth hormone. *Science* **179**, 77–9.

Brobeck, J. R., Tepperman, J., and Long, C. N. H. (1943). Experimental hypothalamic hyperphagia in the albino rat. *Yale J. Biol. Med.* **15**, 831–53.

Brown, M. R. (1981). Neuropeptides: central nervous system effects on nutrient metabolism. *Diabetologia* **20**, 299–304.

—— Fisher, L. A., Rivier, J., Spiess, J., Rivier, C., and Vale, W. (1982). Corticotropin-releasing factor: effects on the sympathetic nervous system and oxygen consumption. *Life Sci.* **30**, 207–10.

—— —— Spiess, J., Rivier, J., Rivier, C., and Vale, W. (1982). Comparison of the biologic actions of corticotropin releasing factor and sauvagine. *Regul. Peptides* **4**, 107–14.

—— Rivier, J., and Vale, W. (1977*a*). Bombesin: potent effects on thermoregulation in the rat. *Science* **196**, 998–1000.

—— —— and Vale, W. (1977*b*). Bombesin affects the central nervous system to produce hyperglycemia in rats. *Life Sci.* **21**, 1729–34.

—— Taché, Y., and Fisher, D. (1979). Central nervous system action of bombesin: mechanism to induce hyperglycemia. *Endocrinology* **105**, 660–5.

Buchan, A. M. J., Polak, J. M., Sullivan, S., Bloom, S. R., Brown, M., and Pearse, A. G. E. (1978). Neurotensin in the gut. In *Gut hormones* (ed. S. R. Bloom), pp. 544–9. Churchill Livingstone, Edinburgh.

Buffa, R., Solcia, E., and Go, V. L. W. (1976). Immunohistochemical identification of the cholecystokinin cell in the intestinal mucosa. *Gastroenterology* **70**, 528–532.

Cantalamessa, F., de Caro, G. and Perfumi, M. (1981). Water intake inhibition and vasopressin release following intra-cerebroventricular administration of tachykinins to rats of the Wistar and Brattleboro strain. *Pharmac. Res. Commun.* **13**, 641–55.

—— —— Massi, M., and Micossi, L. G. (1982). A study of behavioural alterations induced by intracerebroventricular administration of bombesin to rats. *Pharmac. Res. Commun.* **14**, 163–73.

Carraway, R. and Leeman, S. E. (1975). The amino acid sequence of a hypothalamic peptide, neurotensin. *J. Biol. Chem.* **250**, 1907–11.

Chang, M. M. and Leeman, S. E. (1970). Isolation of a sialogogic peptide from bovine hypothalamic tissue and its characterization as substance P. *J. Biol. Chem.* **245**, 4784–90.

—— —— and Naill, H. D. (1971). Amino-acid sequence of substance P. *Nature New Biol.* **232**, 86–7.

Chey, W. Y., Sivasomboon, B., and Hendricks, J. (1973). Actions and interactions of gut hormones and histamine on gastric secretion of acid in the rat. *Am. J. Physiol.* **224**, 852–6.

Clineschmidt, B. V. and McGuffin, J. C. (1977). Neurotensin administered intracisternally inhibits responsiveness of mice to noxious stimuli. *Eur. J. Pharmac.* **46**, 395–6.

Cooper, J. R., Bloom, F. E. and Roth, R. H. (1982). *The biochemical basis of neuropharmacology*. 4th edn. Oxford University Press, New York.

Cooper, S. J. and Sanger, D. J. (1984). Endorphinergic mechanisms in food, salt and water intake: an overview. *Appetite* **5**, 1–6.

Costa, E. and Trabucchi, M. (eds.). (1980). *Neural peptides and neuronal communication. Advances in biochemical psychopharmacology*, Vol. 22. Raven Press, New York.

Costa, M., Furness, J. B., Franco, R., Llewellyn-Smith, I., Murphy, R., and Beardsley, A. M. (1982). Substance P in nerve tissue in the gut. In *Substance P in the nervous system*, Ciba Foundation symposium 91, pp. 129–37. Pitman, London.

Crawley, J. N., Hays, S. E., Paul, S. M., and Goodwin, F. K. (1981). Cholecystokinin reduces exploratory behaviour in mice. *Physiol. Behav.* **27**, 407–11.

Cuello, A. C. and Kanazawa, I. (1978). The distribution of substance P-immunoreactive fibres in the rat central nervous system. *J. Comp. Neurol.* **178**, 129–56.

—— del Fiacco, M., and Paxinos, G. (1978). The central and peripheral ends of the substance P-containing sensory neurones in the rat trigeminal system. *Brain Res.* **152**, 499–509.

Debas, H. T., Farrooq, O., and Grossman, M. I. (1975). Inhibition of gastric emptying is a physiological action of cholecystokinin. *Gastroenterology* **68**, 1211–17.

de Caro, G., Massi, M. and Micossi, L. G. (1978*a*). Antidipsogenic effect of intracranial injections of substance P in rats. *J. Physiol. (Lond.)* **279**, 133–40.

—— —— —— (1978*b*). Effects of eledoisin, physalaemin and some related peptides on water intake and arterial bolld pressure in conscoius rats. *Pharmac. Res. Commun.* **10**, 633–42.

—— —— —— (1978*c*). Potent dipsogenic effect of physalaemin in the pigeon. *Pharmac. Res. Commun.* **10**, 861–6.

—— —— —— and Venturi, F. (1978*d*). Physalaemin, a new potent antidipsogen in the rat. *Neuropharmacology* **17**, 925–9.

—— —— —— (1979). Effect of eledoisin, physalaemin and some eledoisin- or physalaemin-like peptides on water intake and arterial blood pressure in conscious pigeons. *Pharmac. Res. Commun.* **11**, 891–901.

—— —— —— (1980*a*). Modifications of drinking behaviour and of arterial blood pressure induced by tachykinins in rats and pigeons. *Psychopharmacology* **68**, 243–47.

—— —— —— (1980*b*). Bombesin potently stimulates water intake in the pigeon. *Neuropharmacology* **19**, 867–70.

—— Mariotti, M., Massi, M., and Micossi, L. G. (1980*c*). Dipsogenic effect of angiotensin, bombesin and tachykinins in the duck. *Pharmac. Biochem. Behav.* **13**, 229–33.

—— —— —— —— (1980*d*). Relative dipsogenic potency of some partial sequences of bombesin in pigeons and ducks. *Pharmac. Res. Commun.* **12**, 483–7.

—— Massi, M., and Micossi, L. G. (1980*e*). Effect of bombesin on drinking induced by angiotensin II, carbachol and water deprivation in the rat. *Pharmac. Res. Commun.* **12**, 657–66.

—— —— —— and Perfumi, M. (1982). Angiotensin II antagonists versus drinking induced by bombesin or eledoisin in pigeons. *Peptides* **3**, 631–6.

Della-Fera, M. A. and Baile, C. A. (1979). Cholecystokinin octapeptide: continuous picomole injections into the cerebral ventricles of sheep suppress feeding. *Science* **206**, 471–3.

—— —— (1980). CCK-octapeptide injected in CSF decreases meal size and daily food intake in sheep. *Peptides* **1**, 51–54.

—— —— and Peikin, S. R. (1981). Feeding elicited by injection of the cholecystokinin antagonist dibutyryl cyclic GMP into the cerebral ventricles of sheep. *Physiol. Behav.* **26**, 799–801.

De Jong, A., Strubbe, J. M., and Steffens, A. B. (1977). Hypothalamic influence on insulin and glucagon release in the rat. *Am. J. Physiol.* **233**, E380–E388.

Denbow, D. M. and Myers, R. D. (1982). Eating, drinking and temperature responses to intracerebroventricular cholecystokinin in the chick. *Peptides* **3**, 739–43.

Deutsch, J. A. and Hardy, W. T. (1977). Cholecystokinin produces bait shyness in rats. *Nature* **266**, 196.

—— and Parsons, S. L. (1981). Bombesin produces taste aversion in rats. *Behav. Neural Biol.* **31**, 110–13.

Dockray, G. J., Vaillant, C., and Hutchinson, J. B. (1981). Immunochemical characterization of peptides in endocrine cells and nerves with particular reference to gastrin and cholecystokinin. In *Cellular basis of chemical messengers in the digestive system* (eds. M. I. Grossman, M. A. B. Brazier, and J. Lechago) pp. 215–30. Academic Press, New York.

—— —— and Walsh, J. H. (1979). The neuronal origin of bombesin-like immunoreactivity in the rat gastrointestinal tract. *Neuroscience* **4**, 1561–1568.

Eison, A. S., Eison, M. S. and Iversen, S. D. (1982). The behavioural effects of a novel substance P analogue following infusion into the ventral tegmental area or substantia nigra of rat brain. *Brain Res.* **238**, 137–52.

Emson, P. C. and Marley, P. D. (1983). Cholecystokinin and vasoactive intestinal polypeptide. In *Neuropeptides* (eds. L. L. Iversen, S. D. Iversen, and S. H. Snyder) pp. 255–306. Plenum Press, New York.

Epstein, A. N. (1982). The physiology of thirst. In *The physiological mechanisms of motivation* (ed. D. W. Pfaff) pp. 165–214. Springer-Verlag, New York.

Erspamer, V. (1980). Peptides of the amphibian skin active on the gut. II. Bombesin-like peptides: isolation, structure and basic functions. In *Gastrointestinal hormones* (ed. G. B. J. Glass) pp. 343–461. Raven Press, New York.

—— The tachykinin peptide family. *Trends Neurosci.* **4**, 267–9.

—— Falconieri Erspamer, G. F., Improta, G., Negri, L., and de Castiglione, R. (1980). Sauvagine, a new polypeptide from *Phyllomedusa sauvagei* skin. *Naunyn Schmiedebergs Arch. Pharmac.* **312**, 265–70.

—— Melchiorri, P., Falconieri Erspamer, G., and Negri, L. (1978). Polypeptides of the amphibian skin active on the gut and their mammalian counterparts. *Adv. Exp. Med. Biol.* **106**, 51–64.

—— —— and Sopranzi, N. (1972). The action of bombesin on the systemic arterial blood pressure of some experimental animals. *Br. J. Pharmac.* **45**, 442–50.

Evered, M. D. (1983). Neuropeptides and thirst. *Prog. Neuro-Psychopharmac. Biol. Psychiat.* **7**, 469–76.

—— Fitzsimons, J. T. and de Caro, G. (1977). Drinking behaviour induced by intracranial injections of eledoisin and substance P in the pigeon. *Nature* **268**, 332–3.

Falasco, J. D., Smith, G. P., and Gibbs, J. (1979). Cholecystokinin suppresses sham-feeding in the rhesus monkey. *Physiol. Behav.* **23**, 887–90.

Fitzsimons, J. T. (1979). *The physiology of thirst and sodium appetite.* Cambridge University Press, Cambridge.

—— (1980). Angiotensin and other peptides in the control of water and sodium intake. *Proc. R. Soc. Lond.* B. **210**, 164–82.

—— and Evered, M. D. (1978). Elediosin, substance P and related peptides: intracranial dipsogens in the pigeon and antidipsogens in the rat. *Brain Res.* **150**, 533–42.

Frame, C. M., Davidson, M. B., and Sturdevant, R. A. L. (1975). Effects of the octapeptide of cholecystokinin on insulin and glucagon secretion in the dog. *Endocrinology* **97**, 549–53.

Geary, N. and Smith, G.P. (1982*a*). Pancreatic glucagon and postprandial satiety in the rat. *Physiol. Behav.* **28**, 313–22.

—— —— (1982*b*). Pancreatic glucagon fails to inhibit sham feeding in the rat. *Peptides* **1**, 163–6.

—— —— (1983). Selective hepatic vagotomy blocks pancreatic glucagon's satiety effect. *Physiol. Behav.* **31**, 391–4.

Gibbs, J., Falasco, J. D., and McHugh, P. R. (1976). Cholecystokinin-decreased food intake in rhesus monkeys. *Am. J. Physiol.* **230**, 15–18.

—— Fauser, D. J., Rowe, E. A., Rolls, B. J., Rolls, E. T., and Maddison, S. P. (1979). Bombesin suppresses feeding in rats. *Nature* **282**, 208–10.

—— Kulkosky, P. J., and Smith, G. P. (1981). Effects of peripheral and central bombesin on feeding behaviour of rats. *Peptides* **2**, Suppl. 2, 179–83.

—— Young, R. C., and Smith, G. P. (1973*a*). Cholecystokinin decreases food intake in rats. *J. Comp. Physiol. Psychol.* **84**, 488–95.

—— —— and Smith, G. P. (1973*b*). Cholecystokinin elicits satiety in rats with open gastric fistulas. *Nature* **245**, 323–5.

Goriya, Y., Bahoric, A., Marliss, E. B., Zinman, B., and Albisser, A. M. (1981). Diurnal metabolic and hormonal responses to mixed meals in healthy dogs. *Am. J. Physiol.* **240**, E54–E59.

Gosnell, B. A., Morley, J. E., and Levine, A. S. (1983). A comparison of the effects of corticotropin releasing factor and sauvagine on food intake. *Pharmac. Biochem. Behav.* **19**, 771–5.

Greenway, F. L., and Bray, G. A. (1977). Cholecystokinin and satiety. *Life Sci.* **21**, 769–72.

Hanley, M. R., Sandberg, B. E. B., Lee, C. M., Iversen, L. L., Brundish, D. E., and Wade, R. (1980). Specific binding of ^{3}H-substance P to rat brain membranes. *Nature* **286**, 810–12.

Harper, A. A., and Raper, H. S. (1943). Pancreozymin, a stimulant of the secretion of pancreatic enzymes in extracts of the small intestine. *J. Physiol.* **102**, 115–25.

Hays, S. E., Beinfeld, M. C., Jensen, R. I., Goodwin, F. D., and Paul, S. M. (1980). Demonstration of a putative receptor site for cholecystokinin in rat brain. *Neuropeptides* **1**, 53–62.

—— Goodwin, F. K., and Paul, S. M. (1981). Cholecystokinin receptors are decreased in basal ganglia and cerebral cortex of Huntingdon's disease. *Brain Res.* **225**, 452–6.

Hoebel, B. G., Hernandez, L., McLean, S., Stanley, B. G., Aulissi, E. F., Glimcher, P., and Margolin, D. (1982). Catecholamines, enkephalin and neurotensin in feeding and reward. In *The neural basis of feeding and reward* (eds. B. G. Hoebel and D. Novin) pp. 465–78. Haer Institute, Brunswick, Maine.

Hökfelt, T., Efendic, S., Hellerstom, Johansson, O., Luft, R., and Arimura, A. (1975). Cellular localization of somatostatin in endocrine-like cells and neurones of the rat with special reference to the A_1 cells of the pancreatic islets and the hypothalamus. *Acta Endocrinol.* **80**, Suppl. 200, 5–41.

Hökfelt, T., Elfvin, L. G., Schultzberg, M., Goldstein, M., and Nilsson, G. (1977). On the occurrence of substance P-containing fibres in sympathetic ganlia: immunohistochemical evidence. *Brain Res.* **123**, 29–41.

—— Kellerth, J-O., Nilsson, G., and Pernow, B. (1975) Experimental immunohistochemical studies on the localization and distribution of substance P in cat primary sensory neurones. *Brain Res.* **100**, 235–52.

—— Rehfeld, J. F., Skirboll, L., Ivemark, B., Goldstein, M., and Markey, K.

(1980). Evidence for coexistence of dopamine and CCK in mesolimbic neurones. *Nature* **285**, 476–8.

Holt, J., Antin, J., Gibbs, J., Young, R. C., and Smith, G. P. (1974). Cholecystokinin does not produce bait shyness in rats. *Physiol. Behav.* **12**, 497–8.

Houpt, K. A., and Houpt, T. R. (1979). Gastric emptying and cholecystokinin in the control of food intake in suckling rats. *Physiol. Behav.* **23**, 925–9.

Houpt, T. R. (1983). The sites of action of cholecystokinin in decreasing meal size in pigs. *Physiol. Behav.* **31**, 693–8.

—— Anika, S. M., and Wolff, N. C. (1978). Satiety effects of cholecystokinin and caerulein in rabbits. *Am. J. Physiol.* **235**, R23–R28.

Hsiao, S., Wang, C. H. and Schallert, T. (1979). Cholecystokinin, meal pattern and the intermeal interval: can eating be stopped before it starts? *Physiol. Behav.* **23**, 909–14.

Iversen, L. L. (1983). Nonopioid neuropeptides in mammalian CNS. *Ann. Rev. Pharmac. Toxicol.* **23**, 1–27.

—— Iversen S. D. and Snyder, S. H. (eds.). (1983). *Neuropeptides, Handbook of psychopharmacology,* Vol. 16. Plenum Press, New York.

Iversen, S. D. (1982). Behavioural effects of substance P through dopaminergic pathways in the brain. In *Substance P in the nervous system.* Ciba Foundation symposium 91, pp. 307–319. Pitman, London.

Ivy, A. C. and Janacek, H. M. (1959). Assay of Jorpes-Mutt secretin and cholecystokinin. *Acta physiol. scand.* **45**, 220–30.

—— and Oldberg, E. (1928). A hormone mechanism for gallbladder contraction and evacuation. *Am. J. Physiol.* **86**, 599–613.

Jessell, T. M. (1983). Substance P in the nervous system. In *Neuropeptides* (eds. L. L. Iversen, S. D. Iversen and S. H. Snyder) pp.1–105. Plenum Press, New York.

Jorpes, E. and Mutt, V. (1966). Cholecystokinin and pancreozymin one single hormone? *Acta physiol. scand.* **66**, 196–202.

Katz, D. M. and Karten, H. J. (1980). Substance P in the vagal sensory ganglia: localization in cell bodies and pericellular arborizations. *J. Comp. Neurol.* **193**, 549–64.

Katz, R. (1980). Grooming elicited by intracerebroventricular bombesin and eledoisin in the mouse. *Neuropharmacology* **19**, 143–6.

Kelley, A. E., Stinus, L., and Iversen, S. D. (1979). Behavioural activation induced in the rat by substance P infusion into ventral tegmental area; implications of dopaminergic A10 neurones. *Neurosci. Lett.* **11**, 335–9.

Kissileff, H. R., Pi-Sunyer, F. X., Thornton, J., and Smith, G. P. (1981). C-terminal octapeptide of cholecystokinin decreases food intake in man. *Am. J. Clin. Nutr.* **34**, 154–160.

Kobayashi, R. M., Brown, M. R., and Vale, W. (1977). Regional distribution of neurotensin and somatostatin in rat brain. *Brain Res.* **126**, 584–8.

Konturek, S. J. (1980). Somatostatin and opiate peptides: their action on gastrointestinal secretion. In *Gastrointestinal hormones.* (ed. G. B. J. Glass) pp. 693–716. Raven Press, New York.

Kraly, F. S., Carty, W. J., Resmick, S., and Smith, G. P. (1978). Effect of cholecystokinin on meal size and intermeal interval in the sham-feeding rat. *J. Comp. Physiol. Psychol.* **92**, 697–707.

—— Miller, L. A., and Gibbs, J. (1983). Diurnal variation for inhibition of eating by bombesin in the rat. *Physiol. Behav.* **31**, 395–9.

Krulich, L. and McCann, S. M. (1969). Effect of GH-releasing factor and GH-inhibiting factor on the release and concentration of GH in pituitaries incubated in vitro. *Endocrinology* **85**, 319–24.

—— Dhariwal, A. P. S., and McCann, S. M. (1968). Stimulatory and inhibitory effects of purified hypothalamic extracts on growth hormone release from rat pituitary in vitro. *Endocrinology* **83**, 783–90.

Kulkosky, P. J., Breckenridge, C., Krinsky, R., and Woods, S. C. (1976). Satiety elicited by the C-terminal octapeptide of cholecystokinin-pancreozymin in normal and VMH-lesioned rats. *Behav. Biol.* **18**, 227–34.

—— Gibbs, J., and Smith, G. P. (1982). Behavioural effects of bombesin administration in rats. *Physiol. Behav.* **28**, 505–12.

—— Gray, L., Gibbs, J., and Smith, G. P. (1981). Feeding and selection of saccharin, after injections of bombesin, LiCl, and NaCl. *Peptides* **2**, 61–4.

Lang, R. E., Unger, T., Rascher, W. and Ganten, D. (1983). Brain angiotensin. In *Neuropeptides* (eds. L. L. Iversen, S. D. Iversen and S. H. Snyder) pp. 307–61. Plenum Press, New York.

Lee, C. M., Iversen, L. L., Hanley, M. R., and Sandberg, B. E. B. (1982). The possible existence of multiple receptors for substance P. *Naunyn-Schmiedebergs Arch. Pharmac.* **318**, 281–7.

Lembeck, F. (1953). Zur Frage der zentralen Übertragung afferenter Impulse. III. Mittelung. Das Vorkommen und die Bedeutung der Substanz P in den dorsalen Wurzein des Rückenmarkes. *Nauny-Schmiedebergs Arch. Exp. Path. Pharmk.* **238**, 542–5.

Levine, A. S. and Morley, J. E. (1981). Peptidergic control on insulin-induced feeding. *Peptides* **2**, 261–4.

—— —— (1982). Peripherally administered somatostatin reduces feeding by a vagal mediated mechanism. *Pharmac. Biochem. Behav.* **16**, 897–902.

—— —— Kneip, J., Grace, M., and Morley, J. E. (1983). Effect of centrally administered neurotensin on multiple feeding paradigms. *Pharmac. Biochem. Behav.* **18**, 19–23.

—— Rogers, B., Kneip, J., Grace, M., and Morley, J. E. (1983). Effect of centrally administered corticotropin releasing factor (CRF) on multiple feeding paradigms. *Neuropharmacology* **22**, 337–9.

Liebling, D. S., Eisner, J. D., Gibbs, J., and Smith, G. P. (1975). Intestinal satiety in rats. *J. Comp. Physiol.* **89**, 955–65.

Ljungdahl, A., Hökfelt, T., and Nilsson, G. (1978). Distribution of substance P-like immunoreactivity in the central nervous system of the rat. I. Cell bodies and nerve terminals. *Neuroscience* **3**, 861–943.

Lorenz, D. N. and Goldman, S. A. (1982). Vagal mediation of the cholecystokinin satiety effect in rats. *Physiol. Behav.* **29**, 599–604.

—— Kreielsheimer, G., and Smith, G. P. (1979). Effect of cholecystokinin, gastrin, secretion and GIP on sham-feeding in the rat. *Physiol Behav.* **23**, 1065–72.

Lotter, E. C., Krinsky, R., McKay, J. M., Treneer, C. M., Porter, D., and Woods, S. C. (1981). Somatostatin decreases food intake of rats and baboons. *J. Comp. Physiol. Psychol.* **95**, 278–87.

Luttinger, D., King, R. A., Sheppard, D., Strupp, J., Nermeroff, C. B., and Prange, A. J. (1982). The effect of neurotensin on food consumption in the rat. *Eur. J. Pharmac.* **81**, 499–503.

McCaleb, M. L. and Myers, R. D. (1980). Cholecystokinin acts on the hypothalamic 'noradrenergic system' involved in feeding. *Peptides* **1**, 47–9.

McDonald, T. J., Jörnvall, H., Nilsson, G., Vagne, M., Ghatei, M., Bloom, S. R., and Mutt, V. (1979). Characterization of a gastrin-releasing peptide from porcine non-antral gastric tissue. *Biochem. Biophys. Res. Commun.* **90**, 227–33.

McLaughlin, C. L. and Baile, C. A. (1979). Cholecystokinin, amphetamine and diazepam and feeding in lean and obese Zucker rats. *Pharmac. Biochem. Behav.* **10**, 87–93.

—— —— (1980*a*). Feeding response of weanling Zucker obese rats to cholecystokinin and bombesin. *Physiol. Behav.* **25**, 341–6.

—— —— (1980*b*). Decreased sensitivity of Zucker obese rats to the putative satiety agent cholecystokinin. *Physiol. Behav.* **25**, 543–8.

—— —— (1981). Obese mice and the satiety effects of cholecystokinin, bombesin and pancreatic polypeptide. *Physiol. Behav.* **26**, 433–7.

—— Peikin, S. R., and Baile, C. A. (1982). Decreased pancreatic exocrine response to cholecystokinin in Zucker rats. *Am. J. Physiol.* **242**, G612–G619.

—— —— —— (1983*a*). Food intake response to modulation of secretion of cholecystokinin in Zucker rats. *Am. J. Physiol.* **244**, R676–R685.

—— —— —— (1983*b*). Feeding behaviour response of Zucker rats to proglumide, a CCK receptor antagonist. *Pharmac. Biochem. Behav.* **18**, 879–83.

Maddison, S. (1977). Intraperitoneal and intracranial cholecystokinin depress operant responding for food. *Physiol. Behav.* **19**, 819–24.

Mansbach, R. S. and Lorenz, D. N. (1983). Cholecystokinin (CCK-8) elicits prandial sleep in rats. *Physiol. Behav.* **30**, 179–83.

Martin, C. F. and Gibbs, J. (1980). Bombesin elicits satiety in sham-feeding rats. *Peptides* **1**, 131–4.

Martin, J. R. and Novin, D. (1977). Decreased feeding in rats following hepatic-portal infusions of glucagon. *Physiol. Behav.* **19**, 461–6.

—— —— and VanderWeele, D. (1978). Loss of glucagon suppression of feeding after vagotomy in rats. *Am. J. Physiol.* **234**, E314–E318.

Martin, R., Geis, R., Holl, R., Schäfer, M., and Voigt, K. H. (1983). Co-existence of unrelated peptides in oxytocin and vasopressin terminals of rat neurohypophyses: immunoreactive methionine^5- enkephalin-, leucine^5-enkephalin- and cholecystokinin-like substances. *Neuroscience* **8**, 213–27.

Mason, G. A., Nemeroff, C. B., Luttinger, D., Hatley, O. L., and Prange, A. J. (1980). Neurotensin and bombesin: differential effects on body temperature of mice after intracisternal administration. *Regul. Peptides* **1**, 53–60.

Meyer, J. H. and Grossman, M. (1972). Comparison of D- and L-phenylalanine as pancreatic stimulants. *Am. J. Physiol.* **222**, 1058–63.

Miller, R. J. (1983). The enkephalins. In *Neuropeptides* (eds. L. L. Iversen, S. D. Iversen, and S. H. Snyder) pp. 107–207. Plenum Press, New York.

Mineka, S. and Snowdon, C. T. (1978). Inconsistency and possible habituation of CCK-induced satiety. *Physiol. Behav.* **21**, 65–72.

Montecucchi, P. C. and Henschen, A. (1981). Amino acid composition and sequence analysis of sauvagine, a new active peptide from the skin of *Phyllomedusa sauvagei*. *Int. J. Peptide Protein Res.* **18**, 113–20.

Moody, A. J. Jacobsen, H., and Sundby, F. (1978). Gastric glucagon and gut glucagon-like immunoreactants. In *Gut hormones* (ed. S. R. Bloom) pp. 369–78. Churchill Livingstone, Edinburgh.

Moody, T. W. and Pert, C. B. (1979). Bombesin-like peptides in rat brain: quantitation and biochemical characterization. *Biochem. Biophys. Res. Commun.* **90**, 7–14.

—— —— and Jacobowitz, D. M. (1979). Bombesin-like peptides: regional distribution in rat brain. In *Peptides – structure and biological function* (eds. E. Gross and J. Meienhofer) pp. 865–868. Pierce Chem. Co., Rockford, Illinois.

—— —— Rivier, J., and Brown, M. R. (1978). Bombesin: specific binding to rat brain membranes. *Proc. Natn. Acad. Sci. USA* **75**, 5372–6.

Moos, A. B., McLaughlin, C. L., and Baile, C. A. (1982). Effects of CCK on gastrointestinal function in lean and obese Zucker rats. *Peptides* **3**, 619–22.

Moran, T. H. and McHugh, P. R. (1982). Cholecystokinin suppresses food intake by inhibiting gastric emptying. *Am. J. Physiol.* **242**, R491–R497.

Morley, J. E. (1979). Extrahypothalamic thyrotropin releasing hormone – its distribution and functions. *Life Sci.* **25**, 1539–50.

—— (1982). Minireview. The ascent of cholecystokinin (CCK) from gut to brain. *Life Sci.* **30**, 479–93.

—— and Levine, A. S. (1980). Thyrotropin releasing hormone (TRH) suppresses stress induced eating. *Life Sci.* **27**, 269–74.

—— —— (1981). Bombesin inhibits stress-induced eating. *Pharmac. Biochem. Behav.* **14**, 149–51.

—— —— (1982). Corticotrophin releasing factor, grooming and ingestive behaviour. *Life Sci.* **31**, 1459–64.

—— —— and Grace, M., and Kneip, J. (1982*a*). Dynorphin-(1–13), dopamine and feeding in rats. *Pharmac. Biochem. Behav.* **16**, 701–5.

—— —— and Kneip, J. (1981). Muscimol induced feeding: a model to study the hypothalamic regulation of appetite. *Life Sci.* **29**, 1213–18.

—— —— —— and Grace, M. (1982*b*). The effect of vagotomy on the satiety effects of neuropeptides and naloxone. *Life Sci.* **30**, 1943–47.

—— —— and Murray, S. S. (1981). Flavour modulates the antidipsogenic effect of substance P. *Brain Res.* **226**, 334–8.

—— —— —— —— and Kneip, J. (1982*c*). Peptidergic regulation of norepinephrine induced feeding. *Pharmac. Biochem. Behav.* **16**, 225–8.

—— —— Plokta, E. D., and Seal. U. S. (1983). The effect of naloxone on feeding and spontaneous locomotion in the wolf. *Physiol. Behav.* **30**, 331–4.

—— —— Yim, G. K., and Lowy, M. T. (1983). Opioid modulation of appetite. *Neurosci. Biobehav. Rev.* **7**, 281–305.

Mueller, K. and Hsiao, S. (1977). Specificity of cholecystokinin satiety effect: reduction of food but not water intake. *Pharmac. Biochem. Behav.* **6**, 643–6.

—— —— (1978). Current status of cholecystokinin as a short term satiety hormone. *Neurosci. Biobehav. Rev.* **2**, 79–87.

—— —— (1979). Consistency of cholecystokinin satiety effects across deprivation levels and motivational states. *Physiol. Behav.* **22**, 809–15.

Mutt, V. (1980). Cholecystokinin: isolation structure and functions. In *Gastrointestinal hormones* (ed. G. B. J. Glass) pp. 169–221. Raven Press, New York.

Myers, R. D. and McCaleb, M. L. (1981). Peripheral and intrahypothalamic cholecystokinin act on the noradrenergic 'feeding circuit' in the rat's diencephalon. *Neuroscience* **6**, 645–55.

Nemeroff, C. B., Luttinger, D., and Prange, A. J. (1983). Neurotensin and bombesin. In *Neuropeptides* (eds. L. L. Iversen, S. D. Iversen, and S. H. Snyder) pp. 363–466. Plenum Press, New York.

—— Osbahr, A. J., Bissette, G., Jahnke, G., Lipton, M. A., and Prange, A. J. (1978). Cholecystokinin inhibits tail pinch-induced eating in rats. *Science* **200**, 793–4.

Palkovits, M. and Brownstein, M. J. (1983). Extrahypothalamic distribution and action of hypothalamic hormones. In *Neuropeptides* (eds. L. L. Iversen, S. D. Iversen, and S. H. Snyder) pp. 467–87. Plenum Press, New York.

Parrott, R. F. and Baldwin, B. A. (1981). Operant feeding and drinking in pigs following intracerebroventricular injection of synthetic cholecystokinin octapeptide. *Physiol Behav.* **26**, 419–22.

—— —— Batt, R. A. L. (1980). The feeding response of obese mice (genotype, ob/ob) and their wild type littermates to cholecystokinin pancreozymin. *Physiol. Behav.* **24**, 751–3.

Penick, S. B. and Hinkle, L. E. (1961). Depression of food intake induced in healthy subjects by glucagon. *New Engl. J. Med.* **264**, 893–7.

—— and Smith, G. P. (1964). The effects of glucagon on food intake and body weight in man. *J. Obes.* **1**, 1–5.

Pernow, B. (1953). Studies on substance P. Purification, occurrence and biological actions. *Acta Physiol. Scand.* **29**, Suppl. 105, pp. 1–90.

Pert, A., Moody, T. W., Pert, C. B., de Wald, L. A., and Rivier, J. (1980). Bombesin: receptor distribution in brain and effects on nociception and locomotor activity. *Brain Res.* **193**, 209–20.

Phillips, M. I. (1980). The central renin-angiotensin system. In *The role of peptides in neuronal function* (eds. J. L. Barker and T. G. Smith) pp. 389–430. Marcel Dekker, New York.

Pi-Sunyer, X., Kissileff, H. R., Thornton, J., and Smith, G. P. (1982). C-terminal octapeptide of cholecystokinin decreases food intake in obese men. *Physiol. Behav.* **29**, 627–30.

Polak, J. M., Bloom, S. R., Hobbs, S., Solcia, E., and Pearse, A. G. E. (1976). Distribution of bombesin-like peptide in human gastrointestinal tract. *Lancet* **i**, 1109–10.

—— Pearse, A. G. E., Bloom, S. R., Buchan, A. M. J., Rayford, P. L., and Thompson, J. C. (1975). Identification of cholecystokinin-secretin cells. *Lancet* **ii**, 1016–17.

Rehfeld, J. F. (1980). Cholecystokinin as satiety signal. *Int. J. Obes.* **5**, 465–9.

Richter, C. P. A. (1922). A behavioristic study of the activity of the rat. *Comp. Psychol. Monographs* **1**, 1–55.

Rioux, F., Quirion, R., St. Pierre, S., Regoli, D., Jolicoeur, F., Belanger, F., and Barbeau, A. (1981). The hypotensive effect of centrally administered neurotensin in rats. *Eur. J. Pharmac.* **69**, 241–7.

Rivier, C., Brownstein, M., Spiess, J., Rivier, J., and Vale, W. (1982). *In vivo* corticotropin-releasing factor-induced secretion of adrenocorticotropin β-endorphin and corticosterone. *Endocrinology* **110**, 272–8.

Rolls, B. J. and Rolls, E. T. (1982). *Thirst*. Cambridge University Press, Cambridge.

Rorstad, O. P., Martin, J. B., and Terry, L. C. (1980). Somatostatin and the nervous system. In *The role of peptides in neuronal function* (eds. J. L. Barker and T. G. Smith) pp. 573–614. Marcel Dekker, New York.

Saito, A., Goldfine, I. D., and Williams, J. A. (1981). Characterization of receptors for cholecystokinin and related peptides in mouse cerebral cortex. *J. Neurochem.* **37**, 483–90.

—— Sankaran, H., Goldfine, I. D., and Williams, J. A. (1980). Cholecystokinin receptors in the brain. Characterization and distribution. *Science* **208**, 115–56.

—— Williams, J. A., and Goldfine, I. D. (1981). Alterations in brain cholecystokinin receptors after fasting. *Nature* **289**, 599–600.

—— —— Waxler, S. H., and Goldfine, I. R. (1982). Alterations of brain cholecystokinin receptors in mice made obese with goldthioglucose. *J. Neurochem.* **39**, 525–8.

Sanger, D. J. (1981). Endorphinergic mechanisms in the control of food and water intake. *Appetite* **2**, 193–208.

—— (1983). Opiates and ingestive behaviour. In *Theory in psychopharmacology* (ed. S. J. Cooper) Vol. 2, pp. 75–113. Academic Press, London.

Saria, A., Wilson, S. P., Molnar, A., Viveros, O. H., and Lembeck, F., (1980). Substance P and opiate-like peptides in human adrenal medulla. *Neurosci. Lett.* **20**, 195–200.

Savory, C. J. and Gentle, M. J. (1980). Intravenous injections of cholecystokinin and caerulein suppress food intake in domestic fowls. *Experientia* **36**, 1191–2.

—— —— (1983). Brain cholecystokinin and satiety in fowls. *Appetite* **4**, 223.

Schallert, T., Pendergrass, M., and Farrar, S. B. (1982). Cholecystokinin-octapeptide effects on eating elicted by 'external' versus 'internal' cues in rats. *Appetite* **3**, 81–90.

Schally, A. V., Dupont, A., Arimura, A., Redding, T. W., Nishi, N., Linthicum, G. L., and Schlesinger, D. H. (1976). Isolation and structure of somatostatin from porcine hypothalami. *Biochemistry* **15**, 509–14.

Schulman, J. L., Carleton, J. L., Whitney, G., and Whitehorn, J. C. (1957).

Effect of glucagon on food intake and body weight in man. *J. Appl. Physiol.* **11**, 419–21.

Skirboll, L., Hökflet, T., Rehfeld, J., Cuello, A. C., and Dockray, G. (1982). Co-existence of substance P and cholecystokinin-like immunoreactivity in neurones of the mesencephalic periaqueductal central gray. *Neurosci. Lett.* **28**, 35–9.

Smith, G. P. and Gibbs, J. (1979). Postprandial satiety. In *Progress in psychobiology and physiological psychology* (eds. J. M. Sprague and A. N. Epstein) Vol. 8, pp. 179–242. Academic Press, New York.

—— Jerome, C., Cushin, B. J., Eterno. R., and Simansky, K. J. (1981). Abdominal vagotomy blocks the satiety effect of cholecystokinin in the rat. *Science* **213**, 1036–7.

—— Jerome, C., and Gibbs, J. (1981). Abdominal vagotomy does not block the satiety effect of bombesin in the rat. *Peptides* **2**, 409–11.

Sneider, B. S., Monahan, J. W., and Hirsch, J. (1979). Brain cholecystokinin and nutritional status in rats and mice. *J. Clin. Invest.* **64**, 1348–56.

Spiess, J., Rivier, J., Rivier, C., and Vale, W. (1981). Primary structure of corticotropin-releasing factor from ovine hypothalamus. *Proc. Natn. Acad. Sci., USA* **78**, 6517–21.

Stacher, G., Bauer, H., and Steinringer, H. (1979). Cholecystokinin decreases appetite and activation evoked by stimuli arising from the preparation of a meal in man. *Physiol. Behav.* **23**, 325–31.

Stein, L. J. and Woods, S. C. (1981). Cholecystokinin and bombesin act independently to decrease food intake in the rat. *Peptides* **2**, 431–6.

—— —— (1982). Gastrin releasing peptide reduces meal size in rats. *Peptides* **3**, 833–5.

Stinus, L., Kelley, A., and Iversen, S. D. (1978). Increased spontaneous activity following substance P infusion into A10 dopaminergic area. *Nature* **276**, 616–7.

Straus, E. and Yalow, R. S. (1979). Cholecystokinin in the brains of obese and non-obese mice. *Science* **203**, 68–9.

—— —— (1980). Brain cholecystokinin in fasted and fed mice. *Life Sci.* **26**, 969–70.

Stuckey, J. A. and Gibbs, J. (1981). Lateral hypothalamic injections of bombesin suppress food intake in rats. *Soc. Neurosci. Abstr.* **7**, 852.

Sturdevant, R. A. L. and Goetz, H. (1976). Cholecystokinin both stimulates and inhibits human food intake. *Nature* **261**, 713–15.

Swerdlow, N. R., van der Kooy, D., Koob, G. F., and Wenger, J. R. (1983). Cholecystokinin produces conditioned place-aversions, not place preferences, in food-deprived rats: evidence against involvement in satiety. *Life Sci.* **32**, 2087–93.

Unger, R. H. and Orci, L. (1976). Physiology and pathophysiology of glucagon. *Physiol. Rev.* **56**, 778–826.

Vale, W., Spiess, J., Rivier, C., and Rivier, J. (1981). Characterization of a 41-residue ovine hypothalamic peptide that stimulates secretion of corticotropin and β-endorphin. *Science* **213**, 1394–7.

Vanderhaeghen, J. J., Lotstra, F., Vandersande, F., and Dierickx, K. (1981). Coexistence of cholecystokinin and oxytocin-neurophysin in some magnocellular hypothalamo-hypophyseal neurones. *Cell. Tissue Res.* **221**, 227–31.

VanderWeele, D. A. (1982). CCK, endogenous insulin condition and satiety in free-fed rats. *Physiol. Behav.* **29**, 961–4.

—— Geiselman, P. J., and Novin, D. (1979). Pancreatic glucagon, food deprivation and feeding in intact and vagotomized rabbits. *Physiol. Behav.* **23**, 155–8.

—— Pi-Sunyer, F. X., Novin, D., and Bush, M. J. (1980). Chronic insulin infusion suppresses food ingestion and body weight gain in rats. *Brain Res. Bull.* **5**, Suppl. 4, 4–11.

Vijayan, E. and McCann, S. M. (1977). Suppression of feeding and drinking in rats following intraventricular injection of thyrotropin releasing hormone (TRH). *Endocrinology* **100**, 1727–30.

Villareal, J. A. and Brown, M. R. (1978). Bombesin-like peptide in hypothalamus: chemical and immunological characterization. *Life Sci.* **23**, 2729–34.

Vogel, R. A., Cooper, B. R., Barlow, T. S., Prange, A. J., Mueller, R. A., and Breese, G. R. (1979). Effects of thyrotropin-releasing hormone on locomotor activity, operant performance and ingestive behavior. *J. Pharmac. Exp. Ther.* **208**, 161–9.

Von Euler, U. S. and Gaddum, J. J. (1931). An unidentified depressor substance in certain tissue extracts. *J. Physiol. (Lond.)*. **72**, 74–87.

Walsh, J. H. (1981). Nature of gut peptides and their possible functions. In *Cellular basis of chemical messengers in the digestive system* (eds. M. I. Grossman, M. A. B. Brazier and J. Lechago) pp. 3–11. Academic Press, New York.

—— Wong, H. C., and Dockray, G. J. (1979). Bombesin-like peptides in mammals. *Fed. Proc.* **38**, 2315–19.

West, D. B., Williams, R. H., Braget, D. J., and Woods, S. C. (1982). Bombesin reduces food intake of normal and hypothalamically obese rats and lowers body weight when given chronically. *Peptides* **3**, 61–7.

Wilson, M. C., Denson, D., Bedford, J. A., and Hunsinger, R. N. (1983). Pharmacological manipulation of sincalide (CCK-8)-induced suppression of feeding. *Peptides* **4**, 351–7.

Woods, S. C., West, D. B., Stein, L. J., McKay, L. D., Lotter, E. C., Porte, S. G., Kenney, N. J., and Porte, D. (1981). Peptides and the control of meal size. *Diabetologia* **29**, 305–13.

Young, R. C., Gibbs, J., Antin, J., Holt, J., and Smith, G. P. (1974). Absence of satiety during sham-feeding in the rat. *J. Comp. Physiol. Psychol.* **87**, 795–800.

Young, W. S. and Kuhar, M. J. (1979). Neurotensin receptors: autoradiographic localization in rat CNS. *Eur. J. Pharmac.* **59**, 161–3.

—— —— (1981). Neurotensin receptors localization by light microscopic autoradiography in rat brain. *Brain Res.* **150**, 431–5.

Zarbin, M. A., Innis, R. B., Wamsley, J. K., Snyder, S. H., and Kuhar, M. J. (1981). Autoradiographic localization of CCK receptors in guinea pig brain. *Eur. J. Pharmac.* **71**, 349–50.

Zetler, G. (1980). Effects of cholecystokinin-like peptides on rearing activity and hexobarbital-induced sleep. *Eur. J. Pharmac.* **66**, 137–9.

—— (1981). Central depressant effects of caerulein and cholecystokinin octapeptide (CCK-8) differ from those of diazepam and haloperidol. *Neuropharmacology* **20**, 277–83.

3

Effect of food intake on brain transmitter amine precursors and amine synthesis

G. CURZON

INTRODUCTION

This chapter concerns mainly the acute effects on brain transmitter amine precursors and transmitter amine synthesis of food withdrawal and refeeding. Major dietary constituents (protein and carbohydrate) are considered. Effects of particular items of diet, or chronic effects of diets that are deficient in specific transmitter amine precursors, will not be discussed.

The role of dietary protein is of primary importance here, as the amino acids tyrosine and tryptophan are needed to make brain catecholamines and 5-hydrotryptamine (5HT) respectively. A key point is that brain tryptophan hydroxylase is rate limiting for 5HT synthesis and is normally only about half-saturated with its substrate tryptophan, so that altering the supply of tryptophan to the brain will alter 5HT synthesis. The whole literature on food intake and transmitter amine synthesis starts from the observation that giving tryptophan to rodents increases 5HT synthesis. This leads to a number of questions. Are normal dietary variations of tryptophan (and tyrosine) supply sufficient to alter 5HT (and catecholamine) synthesis? If these amine changes do occur, are there circumstances in which they affect behaviour and other brain functions? In particular, might they influence appetite or food selection? This chapter will only deal systematically with the question of whether normal dietary variations of amino-acid supply affect brain amine synthesis.

For a long time, the idea that the brain was, on the whole, protected from metabolic insult led to the unvoiced assumption that normal dietary differences were hardly likely to affect the synthesis of important brain substances like transmitters. However, there were early indications that large doses of tryptophan increased rat brain 5HT synthesis (Hess and Doepfner, 1961; Eccleston, Ashcroft and Crawford, 1965). Fernstrom and Wurtman (1971*a*) showed that this also occurred on giving relatively small doses approximating to only 10 per cent of optimum daily intake. A group of papers from Wurtman's laboratory followed, on the effects of various dietary manipulations on 5HT synthesis (Fernstrom and Wurtman 1971b, 1972; Colmenares, Wurtman and Fernstrom 1975). Concurrently, my laboratory was depriving rats of food for 24 h, on the assumption that this might decrease the supply of tryptophan to the brain and hence decrease 5HT synthesis. To our surprise, they both increased (Curzon, Joseph and Knott 1972). Another finding that, superficially, seemed just as paradoxical came from Fernstrom and Wurtman (1971*b*). They found that a tryptophan-free high-carbohydrate meal did not decrease brain tryptophan and 5HT synthesis, but increased them. These results are all explicable in terms of the scheme shown in Fig. 3.1.

Thus, Wurtman and his colleagues interpreted the effect of a high-carbohydrate meal in terms of insulin increasing uptake by muscle and hence decreasing plasma concentrations of large neutral amino acids (principally phenylalanine, tyrosine, leucine, isoleucine and valine). These compete with tryptophan and with each other for transport to the brain. Leucine, isoleucine and valine fall much more strikingly than phenylalanine and tyrosine after a carbohydrate meal, while tryptophan falls least of all. As a result, the transport of these aromatic amino acids to the brain is increased.

We explained the effect of food deprivation by the well-known resultant mobilization of fat and increase of plasma unesterified fatty acids. These substances bind to plasma albumin, thus weakening the binding of tryptophan to it and making the freed tryptophan available to the brain (Curzon, Friedel, and Knott 1973; Knott and Curzon, 1972). Wurtman's group have consistently used the ratio of plasma total tryptophan to competing amino acids as an index of the availability of tryptophan to the brain. We find that free tryptophan is a better index (Bloxam and Curzon 1978; Bloxam, Tricklebank, Patel, and Curzon 1980; Gillman, Bartlett, Bridges, Hunt, Patel, Kantamaneni, and Curzon 1981; Sarna, Tricklebank, Kantamaneni, Hunt, Patel, and Curzon 1982) though in certain situations, in particular following food intake, the competers do also have to be taken into account. Recent review articles give the impression that previous apparent disagreements on the above issues are now becoming resolved (Curzon and Sarna 1984; Fernstrom 1983; Sved 1983).

Uptake kinetics is another important variable, but there are a lot of gaps in our knowledge of this. It also must be pointed out that there have been numerous studies on the transport of tryptophan to the brain, but in very few of these (e.g. Sarna *et al.* 1982) have all the major known variables been measured.

Though relatively moderate periods of food deprivation or the ingestion of a carbohydrate-rich tryptophan-free meal increase brain tryptophan, deprivation of dietary tryptophan must eventually deplete tryptophan stores. In the acute

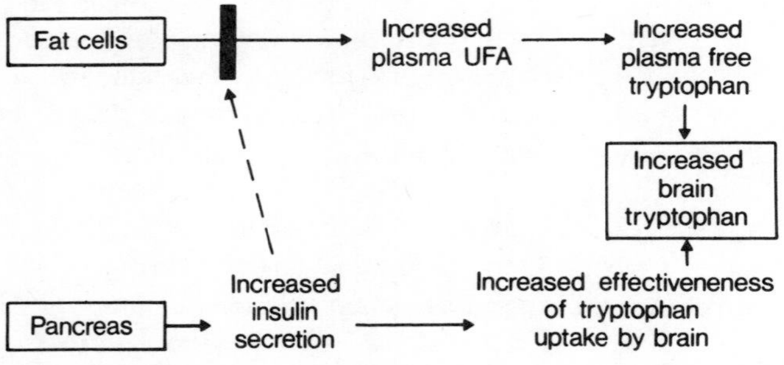

FIG. 3.1 Effects of lipolysis and insulin secretion on rat brain tryptophan (modified from Curzon and Fernando 1976).

situation, it is clear that mechanisms operate which oppose this depletion and which may serve to protect the brain from deficiency of this important amino acid. They also have implications in other dietary circumstances, e.g. during refeeding after food deprivation.

FREE TRYPTOPHAN: REFEEDING AFTER FOOD DEPRIVATION

Not only does moderate food deprivation tend to increase the amount of tryptophan that gets to the brain from the blood but, conversely, refeeding tends to decrease brain tryptophan and thus also to decrease 5HT synthesis. This was first noted a decade ago (Perez-Cruet, Tagliamonte, Tagliamonte, and Gessa 1972; Perez-Cruet, Chase, and Murphy 1974). The magnitude of the effect presumably depends on individual variations in fat mobilization and on the duration of previous food deprivation. It may also be influenced by concurrent changes in competing amino acids in plasma.

Table 3.1 shows the results of a recent investigation of refeeding (Sarna, Kantamaneni, and Curzon 1984). Rats were trained to eat within a 4 h period during the dark (red light). Incidentally, though rats eat mostly during the dark (Le Magen and Devos 1980), almost all reported work on acute effects of feeding on amine precursors involved rats trained to eat during the light period. This is worth pointing out as Larue-Archagiotis and Le Magen (1983) find that fasting increases rat plasma unesterified fatty acid more markedly during the night than during the day.

In this experiment, we see falls of plasma unesterified fatty acid, plasma free tryptophan and brain tryptophan. There is also a rise of plasma competers. The brain change may reflect that of plasma free tryptophan or of its ratio to that of the competers, but it can hardly reflect plasma total tryptophan, as this rises. Nor does it reflect the ratio of total plasma tryptophan to the competers as this

TABLE 3.1 *Effect of standard meal after food withdrawal*

	Percentage change	
	Experiment 1	Experiment 2
Brain tryptophan	−23	nd
CSF tryptophan	nd	−24
Plasma		
Free tryptophan	−27	−29
Total tryptophan	+31	+43
Unesterified fatty acid	−52	nd
Competers	+30	+43
Free tryptophan/competers	−48	−50
Total tryptophan/competers	+1	+1

Experiment 1. 6 rats. Standard chow. 4 h (Sarna *et al.* 1984)
Experiment 2. 6 humans. Hospital meal. 4 h (Perez-Cruet *et al.* 1974)
All changes are significant ($p < 0.01$) except total tryptophan/competers.

remains constant. Table 3.1 also shows that very similar results were obtained in human subjects by Perez-Cruet *et al.* (1974) who used CSF tryptophan as an index of brain tryptophan concentration.

THE ROLE OF COMPETERS

How important then are physiological alterations in plasma competer concentrations? Analysis of the literature suggests they may be less important than has often been implied. Fig. 3.2 summarizes data from three different papers. Each point shows the mean value for a group of rats given a meal of particular composition. The plot of brain tryptophan against the ratio of serum tryptophan to competers for rats given different diets might be taken to indicate a striking influence of the effect of diet on this ratio, and hence on brain tryptophan and

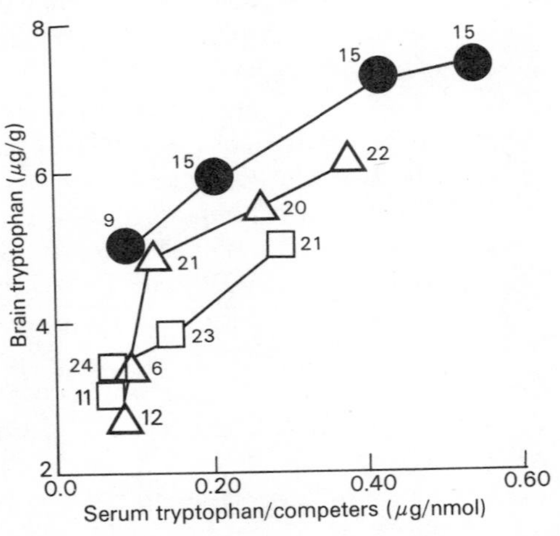

FIG. 3.2 Relationships between rat brain tryptophan concentration and the ratio of serum total tryptophan concentration to the sum of the serum concentrations of amino acids competing with it for transport to the brain. Each symbol indicates the mean value for a group of rats given one meal of a particular composition per 24 h. Numbers against symbols indicate mean serum total tryptophan concentrations (μg/ml). ○, Fernstrom *et al.* (1975); △, Fernstrom *et al.* (1976); □, Fernstrom and Faller (1978).

5HT synthesis. But ratios can be altered not only by changing the denominator (the sum of the concentrations of the competing amino acids) but also by changing the numerator (tryptophan). It is clear that the very steep relationship shown by some of the results from the 1976 paper is completely explicable in terms of serum tryptophan alone and that this also could explain part of the relationship shown by the 1975 paper. If the results on groups of rats with comparable serum tryptophan values are considered alone, it appears that a four-fold range of competer concentrations resulted in a 1.6-fold range of brain tryptophan values for the 1978 study, and in rather smaller responses in the other two studies. These results may be compared with those derived from a study of tryptophan influx into brain from radioactive tryptophan injected into the carotid (Sarna *et al.* 1982). In this work, influx values at different tryptophan concentrations (Fig. 3.3A) were explicable in terms of a saturable (Fig. 3.3B) and an unsaturable component. (Note that effectiveness of uptake decreases with age; this could have important implications, especially in view of recent evidence that plasma tryptophan concentration is low in the elderly). The derived influx equation was used to calculate the effects of different plasma competer concentrations on tryptophan influx. A typical daytime plasma competer level certainly inhibits influx markedly, in comparison with a non-physiological situation - the complete absence of competers. But the effects of

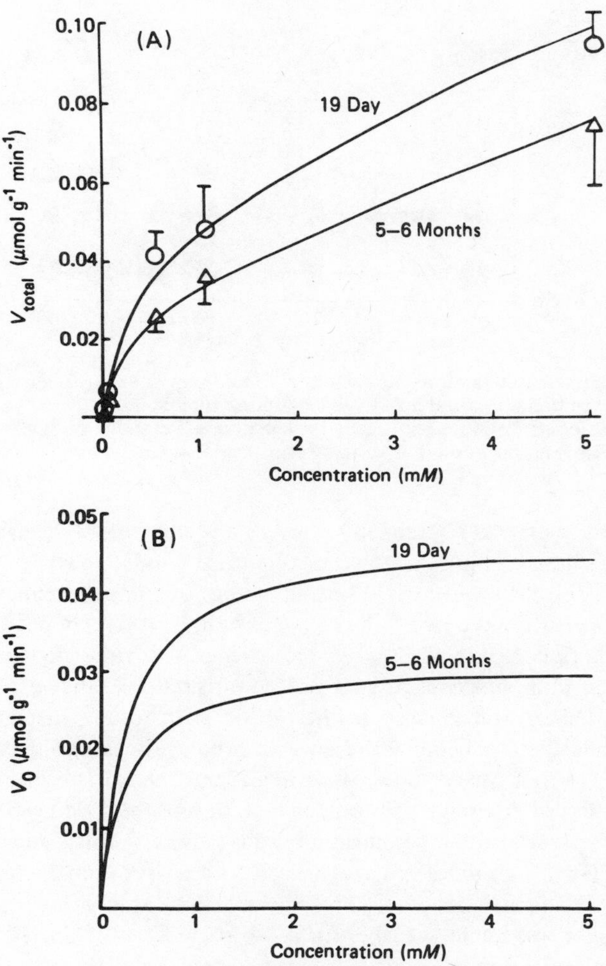

FIG. 3.3 Influx of tryptophan into the brain of pentobarbital anaesthetized rats. (A) V_{total} is the rate of saturable plus nonsaturable influx. Each point is the mean from at least four rats with the SD indicated by vertical bars. Error bars are not shown where they are less than the size of the symbol. (B) V_0 represents the saturable component of influx. The abscissa is the concentration of substrate in the injection solution. From Sarna *et al.* (1982).

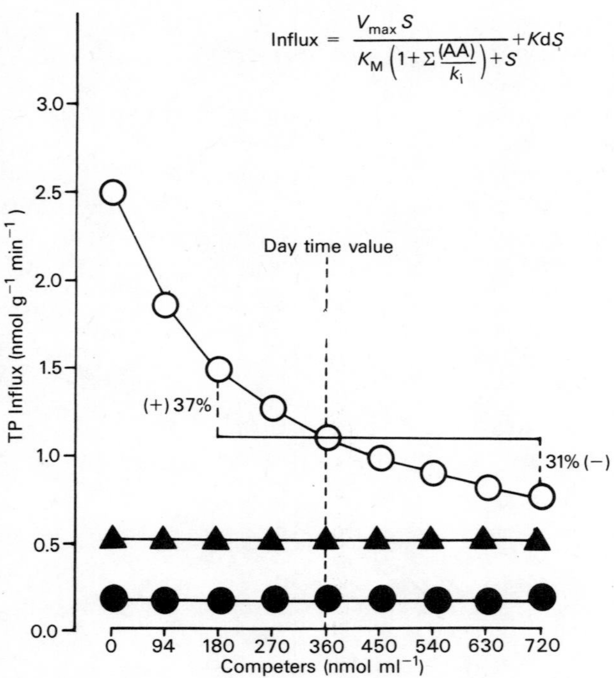

FIG. 3.4 Calculated effects on tryptophan influx into brain of amino acids competing with it for transport to brain. ○, total influx; ♦ and ▲, nonsaturable influx. Values obtained using results by Sarna *et al.* (1982), kinetic data for competers (Pardridge and Oldendorf (1975)) and a tryptophan concentration of 4 μg/ml. Results are comparable over a tryptophan range of 0.4–40 μg/ml.

physiologically more likely competer changes are less striking. Halving a typical daytime value increases influx by about one third. Doubling it, decreases influx by about one third. So a four-fold overall change of competer concentrations is predicted to cause about a two-fold change of brain tryptophan influx.

This prediction agrees well with the effect of a large high-carbohydrate protein-free meal in which we used a procedure similar to that used previously (Fernstrom, Faller, and Shabshelowitz 1975; Fernstrom, Hirsch, and Faller 1976), a standard chow being withdrawn 18 h before making the carbohydrate diet available so that the rat then eats a single large meal. In essential agreement with the earlier work, brain tryptophan rose during the feeding period. There was a concurrent fall of competing amino acids (Fig. 3.5) which our brain influx experiments (Fig. 3.4) predicted would lead to a 57 per cent increase of brain tryptophan. The actual rise was 63 per cent. Thus, when a high-carbohydrate protein-free diet was made available for a 2 h period out of the 24 h, then the large fall of plasma competers led to an increase of brain tryptophan. Note that the time course of the subsequent fall of brain tryptophan almost exactly paralleled that of the ratio of plasma free tryptophan to competers.

This was a rather unusual experiment as neither plasma total nor free tryptophan concentrations altered appreciably during the period of food intake.

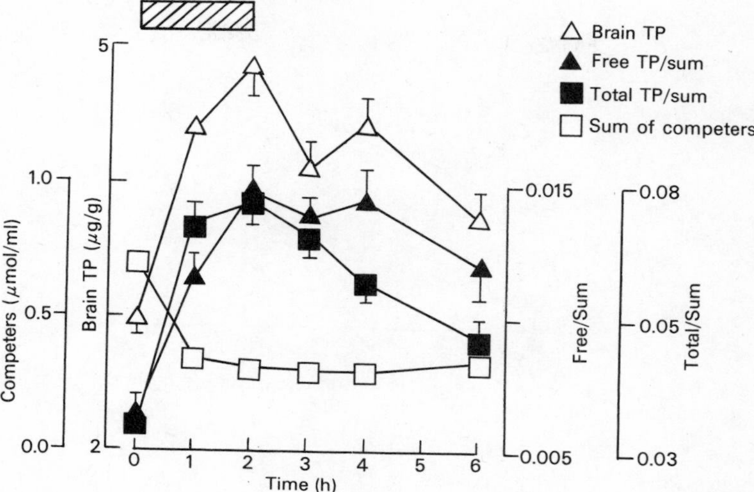

FIG. 3.5 Effect of a high carbohydrate meal at 11.30 h–13.30 h (06.00 h–18.00 h white light) on rat plasma and brain tryptophan and on plasma competing amino acids. Plasma free and total tryptophan concentrations (μg/ml) are divided by the sum of the competers (nmol/ml). 6 rats/group. Values shown + or – SE. ▨ indicates period of exposure to food. (Sarna *et al.* 1984).

However, this permitted a very clear confirmation of the effect of the fall of plasma competers which occurred on feeding carbohydrate.

How a particular meal affects brain tryptophan concentration presumably depends not only on its composition but also on how much brain tryptophan has already been elevated by previous food deprivation. As indicated in Fig. 3.6, the greater this elevation is, then the smaller will be any additional rise due to the effect of a carbohydrate meal on plasma competers as there will be an opposing effect as a result of the fall of plasma free tryptophan due to fatty acid changes. In some circumstances, the latter effect might be sufficient completely to prevent the elevation of tryptophan by carbohydrate. Conversely, if a protein-containing meal is eaten with a composition such that the ratio of plasma tryptophan to competers does not alter markedly, then brain tryptophan will decrease according to the extent to which previous food deprivation has raised plasma free tryptophan concentration.

Figure 3.6 may explain some recent work of Glaeser, Maher and Wurtman (1983) who, unlike Fernstrom and Wurtman (1971*b*) and Fernstrom *et al.* (1975, 1976), found that a protein-free meal after overnight fasting did not increase rat brain tryptophan even though the increased ratio of serum-total tryptophan to competers pointed to a considerable rise. This discrepancy would have been explicable by a fall of free tryptophan on feeding after food deprivation. A 40 per cent casein meal actually decreased brain tryptophan, even though the serum ratio suggested that, if anything, it would rise. Again, this result would be readily explicable by a fall of free tryptophan, This work also illustrates a difference between the control of brain tryptophan and tyrosine

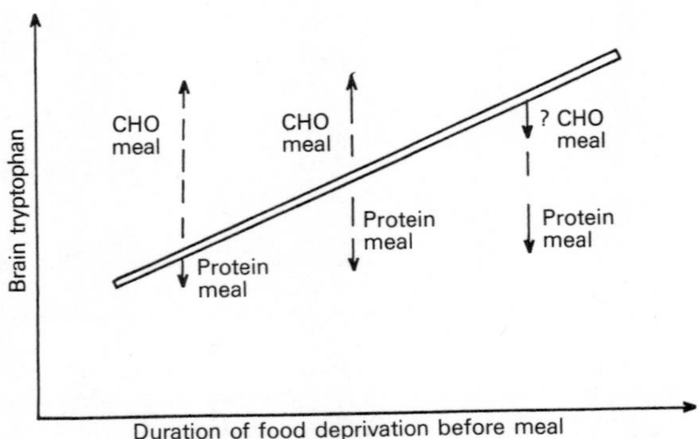

FIG. 3.6 Proposed effects of carbohydrate and protein meals on rat brain tryptophan concentration after different periods of food deprivation.

concentrations. Complications due to much of the plasma amino acid being present in bound form only apply in the case of tryptophan. Therefore, the ratio of serum tyrosine to competers predicts brain tyrosine concentrations very well.

We see then that rat brain tryptophan and tyrosine concentrations can be acutely influenced by diet. However, the most important consequence of this may be that the brain is protected against tryptophan depletion when one might have imagined that there was a danger of this occurring, i.e. during moderate periods of food deprivation or on ingestion of a tryptophan-free meal. Thus, Glaeser *et al.* (1983) obtained identical brain tryptophan concentrations whether their rats were fasted or ate a large protein-free meal and these concentrations were *higher* than those obtained after eating meals containing 18 per cent or 40 per cent casein.

To what extent can the effects of carbohydrate meals in the rat be extrapolted to man? This is by no means clear. Ashley, Barclay, Chauffard, Moennoz, and Leathwood (1982) found that human subjects fed single high-carbohydrate meals or meals containing 20 g protein (which animal work suggests might lead to very different ratios of plasma tryptophan to competers) in fact had similar ratios. On the other hand, Fernstrom, Wurtman, Hammarstrom-Wiklund, Rand, Munro, and Davidson (1979) found that ratios for subjects fed three protein-free meals in a day were twice as large as those for the same subjects fed high-protein meals (150 g/day). Corresponding ratios for tyrosine showed much smaller differences.

TRANSMITTER AMINE SYNTHESIS

5-Hydroxytryptamine

When brain tryptophan concentration does alter on feeding, then 5HT synthesis will be affected, but on the whole to a less marked extent (Fernstrom and Wurtman 1972; Colmenares *et al.* 1975). Whether such changes influence appetite and food selection will be discussed in Chapter 00. Another point of some importance is that the effect of precursor availability on net transmitter metabolism depends on the previous state of repletion of the neurones with transmitter.

This is illustrated by the effect of giving tryptophan to control rats and to animals with 5HT synthesis partly inhibited by *p*-chlorophenylalanine (Marsden and Curzon 1976). In the controls, tryptophan caused a small increase of brain 5HT but its metabolite 5HIAA rose considerably, i.e. a large fraction of the 5HT formed from the tryptophan was rapidly destroyed. In the *p*-chlorophenylalanine-treated rats, the increase of 5HT was greater (even though 5HT synthesis was impaired) and degradation to 5HIAA was less striking. Not only was the metabolism of the injected tryptophan different in the two circumstances but also its effect on behaviour. Thus, the hyperactivity of rats caused by partial inhibition of 5HT synthesis was prevented by tryptophan at doses which did not affect the activity of normal rats in the same test situation.

More work on the effects of feeding during the dark period on transmitter amine metabolism and behaviour would be of interest, as it is in this period that evidence suggests that brain 5HT synthesis is least likely to be able to keep pace with demand (Hutson, Sarna, and Curzon 1984). It is also worth pointing out that Krieger, Crowley, O'Donohue, and Jacobowitz (1980) showed that altering the period during which food is available had markedly different effects on brain 5HT levels in different regions. These effects cannot be simply explained in terms of changes of precursor amino acids and they need further investigation.

Catecholamines

While tryptophan hydroxylase is normally about half saturated with substrate in the rodent brain, tyrosine hydroxylase is much closer to saturation (Carlsson and Lindqvist 1978; Sved 1983). Therefore, catecholamine synthesis is less sensitive than 5HT synthesis to precursor availability. Furthermore, feedback controls impose further limitations (Fernstrom 1983). Results on human ventricular CSF are consistent (Curzon, Kantamaneni, Bartlett, and Bridges 1976) as a significant correlation was found between the concentrations of tryptophan and the 5HT metabolite 5HIAA but not between those of tyrosine and the dopamine metabolite HVA. Whether tyrosine availability, within its normal range, does influence central catecholamine synthesis, is controversial (Biggio, Porceddu, Fratta, and Gessa 1977; Edwards and Rizk, 1981; Badawy and Williams 1982). It may be more sensitive to tyrosine availability when catecholamine neurones are activated (Sved 1983). Indeed, Marcou and Kennett in my laboratory have recently shown that tyrosine elevated CSF dopamine metabolites in stressed rats but not in control animals in a quiet environment.

Evidence that feeding enhances catecholamine turnover independently of tyrosine changes is of considerable interest (Heffner, Hartmann, and Seiden 1980; Biggio *et al.* 1977; Sved 1983), though it is not clear to what extent the changes may reflect associated motor acitvity. Heffner *et al.* reported increased dopamine metabolite concentrations in the nucleus accumbens, amygdala and hypothalamus but not in the striatum, olfactory tubercle, septum or frontal cortex, while Sved found increases in the striatum. In addition, the inhibition of firing of dopamine neurones on giving glucose so that its blood levels were comparable to those on feeding (Saller and Chiodo 1980), must be taken into account.

Much of the previous research effort on the effects of feeding on transmitter synthesis has been done using diets of unusual composition, presented after many hours of food deprivation. There are good tactical reasons for this kind of experimental design and the approach has led to some interesting results. However, the relevance of these results to the range of normal circumstances should not be overestimated. In the future, more emphasis might perhaps be placed on effects of meals of less extreme composition and of different sizes; on the effects of food deprivation and repletion; on regional differences in the effects of feeding and on transmitter changes not related to alterations in the supply of precursors on feeding.

References

Ashley, D. V., Barclay, D. V., Chauffard, F. A., Moennoz, D., and Leathwood, P. D. (1982). Plams amino acid responses in humans to evening meals of differing nutritional composition. *Am. J. Clin. Nutr.* **36**, 142–53.

Badawy, A. A. B. and Williams, D. L. (1982). Enhancement of rat brain catecholamine synthesis by administration of small doses of tyrosine and evidence for substrate inhibition of tyrosine hydroxylase activity by large doses of the amino acid. *Biochem. J.* **206**, 165–8.

Biggio, C., Porceddu, M. L., Fratta, W., and Gessa, G. L. (1977). Changes in dopamine metabolism associated with fasting and satiation. *Adv. Biochem. Psychopharmac.* **16**, 377–83.

Bloxam, D. L. and Curzon, G. (1978). A study of proposed determinants of brain tryptophan concentration in rats after portocaval anastomosis or sham operation. *J. Neurochem.* **31**, 1255–63.

—— Tricklebank, M. D., Patel, A. J., and Curzon, G. (1980). Effects of albumin, amino acids and clofibrate on the uptake of tryptophan by the rat brain. *J. Neurochem.* **34**, 43–9.

Carlsson, A. and Lindqvist, M. (1978). Dependence of 5HT and catecholamine synthesis on concentrations of precursor amino acids in rat brain. *Naunyn-Schmiedebergs Arch. Pharmac.* **303**, 157–64.

Colmenares, J. L., Wurtman, R. J. and Fernstrom, J. D. (1975). Effect of ingestion of a carbohydrate-fat meal on the levels and synthesis of 5-hydroxyindoles in various regions of the rat central nervous system. *J. Neurochem.* **25**, 825–9.

Curzon, G. and Fernando, J. C. R. (1976). Effect of aminophylline on tryptophan and other aromatic amino acids in plasma, brain and other tissues and on brain 5-hydroxytryptamine metabolism. *Br. J. Pharmac.* **58**, 533–45.

– – Sarna, G. (1984). Tryptophan transport to the brain: newer findings and older ones reconsidered. In *Tryptophan: fourth international meeting on tryptophan metabolism, biochemistry, pathology and regulation*, (eds. Schlossberger, H. G., Kochen, W., Linzen, B., and Steinhart, H.) pp. 145–60. DeGruyter, Berlin.

—— Friedel, J., and Knott, P. J. (1973). The effects of fatty acids on the binding of tryptophan to plasma protein. *Nature* **242**, 198–200.

—— Joseph, M. H. and Knott, P. J. (1972). Effects of immobilization and food deprivation on rat brain tryptophan metabolism. *J. Neurochem.* **19**, 1967–74.

—— Kantamaneni, B. D., Bartlett, J. R., and Bridges, P. K. (1976). Transmitter precursors and metabolites in human ventricular cerebrospinal fluid. *J. Neurochem.* **26**, 613–15.

Eccleston, D., Ashcroft, G. W., and Crawford, T. B. B. (1965). 5-Hydroxyindole metabolism in rat. A study of intermediate metabolism using the technique of tryptophan loading. *J. Neurochem.* **12**, 493–503.

Edwards, D. J. and Rizk, M. (1981). Effects of amino acids precursors on catecholamine synthesis in the brain. *Progr. Neuro-Psychopharmac.* **5**, 569–72.

Fernstrom, J. D. (1983). Role of precursor availability in control of monoamine biosynthesis in brain. *Physiol Rev.* **63**, 484–546.

—— and Faller, D. V. (1978). Neutral amino acids in the brain: changes in response to food ingestion. *J. Neurochem.* **30**, 1531–38.

—— and Wurtman, R. J. (1971*a*). Brain serotonin content: physiological dependence on plasma tryptophan levels. *Science* **173**, 149–52.

—— —— (1971*b*). Brain serotonin content: increase following ingestion of carbohydrate diet. *Science* **171**, 1023–25.

—— —— (1972). Brain serotonin content: physiological regulation by plasma neutral amino acids. *Science* **178**, 414–6.

—— Hirsch, M. J., and Faller, D. V. (1976). Tryptophan concentrations in rat brain: failure to correlate with free serum tryptophan or its ratio to the sum of other serum neutral amino acids. *Biochem. J.* **160**, 589–95.

—— Faller, D. V. and Shabshelowitz, H. (1975). Acute reduction of brain serotonin and 5HIAA following food consumption: correlation with the ratio of serum tryptophan to the sum of competing neutral amino acids. *J. Neural Trans.* **36**, 113–21.

—— Wurtman, R. J., Hammarstrom-Wiklund, B., Rand, W. M., Munro, H. N., and Davidson, C. S. (1979). Diurnal variations in plasma concentrations of tryptophan, tyrosine and other neutral amino acids: effect of dietary protein intake. *Am. J. Clin. Nutr.* **32**, 1912–22.

Gillman, P. K., Bartlett, J. R., Bridges, P. K., Hunt, A., Patel. A. J., Kantamaneni, B. D., and Curzon, G. (1981). Indolic substances in plasma, cerebrospinal fluid and frontal cortex of human subjects infused with saline or tryptophan. *J. Neurochem.* **37**, 410–17.

Glaeser, B. S., Maher, T. J., and Wurtman, R. J. (1983). Changes in brain levels of acidic, basic and neutral amino acids after consumption of single meals containing various proportions of protein. *J. Neurochem.* **41**, 1016–21.

Heffner, T. G., Hartman, J. A., and Seiden, L. S. (1980). Feeding increases dopamine metabolism in the rat brain. *Science* **208**, 1168–70.

Hess, S. and Doepfner, W. (1981). Behavioural effects and brain amine content in rats. *Arch. Int. Pharmacodyn.* **134**, 89–99.

Hutson, P. H., Sarna, G. S. and Curzon, G. (1984). Determination of daily variation of brain 5-hydroxytryptamine and dopamine turnovers and of the clearance of their acidic metabolites in conscious rats by repeated sampling of cerebrospinal fluid. *J. Neurochem,* **43**, 291–3.

Knott, P. J. and Curzon, G. (1972). Free tryptophan in plasma and brain tryptophan metabolism. *Nature* **239**, 452–53.

Krieger, D. T., Crowley, W. R., O'Donohue, T. L., and Jacobowitz, D. M. (1980). Effects of food restriction on the periodicity of corticosteriods in plasma and on monoamine concentrations in discrete brain nuclei. *Brain Res.* **188**, 166–74.

Larue-Achagiotis, C. and Le Magen, J. (1983). Fast-induced changes in plasma glucose, insulin and free fatty acid concentration compared in rats during the night and day. *Physiol. Behav.* **30**, 93–6.

Le Magen, J. and Devos, M. (1980). Parameters of the meal pattern in rats: their assessment and physiological significance. *Neurosci. Biobehav. Revs.* **4**, Suppl.1, 1–11.

Marsden, C. A. and Curzon, G. (1976). Studies on the behavioural effects of tryptophan and *p*-chlorophenylalanine. *Neuropharmacology* **15**, 165–71.

Pardridge, W. M. and Oldendorf, W. H. (1975). Kinetic analyses of blood brain barrier transport of amino acids. *Biochim. Biophys. Acta.* **401**, 128–36.

Perez-Cruet, J., Chase, T. N. and Murphy, D. L. (1974). Dietary regulation of brain tryptophan metabolism by plasma ratio of free tryptophan and neutral amino acids in humans. *Nature* **248**, 693–5.

—— Tagliamonte, A., Tagliamonte, P., and Gessa, G. L. (1972). Changes in brain serotonin metabolism associated with fasting and satiation in rats. *Life Sci.* **11**, Part 2, 31–9.

Saller, C. F. and Chiodo, L. A. (1980) Glucose suppresses basal firing and haloperidol-induced increases in the firing rate of central dopaminergic neurones. *Science* **210**, 1269–71.

Sarna, G. S., Kantamaneni, B. D., and Curzon, G. (1984). Variables influencing the effect of a meal on brain tryptophan.*J. Neurochem.* In the press.

—— Tricklebank, M. D., Kantamaneni, B. D., Hunt, A., Patel, A. J., and Curzon, G. (1982). Effect of age on variables influencing the supply of tryptophan to the brain. *J. Neurochem.* **39** 1283–90.

Sved, A. F. (1983). Precursor control of the function of monoaminergic neurrones. In *Nutrition and the brain,* (eds. Wurtman. R. J. and Wurtman, J. J.), Vol 6, pp. 223–75. Raven Press, New York.

4

Psychopharmacology of centrally acting anorectic agents

JOHN E. BLUNDELL

1. SIMPLICITY AND COMPLEXITY: A SYSTEMS VIEW

Psychopharmacological work on centrally-acting anorectic agents has been directed towards two major goals. First, there have been those investigations designed to reveal the central mechanisms involved in the control of food intake, and second, those which have attempted to find safe antiobesity agents that work by suppressing the willingness to eat. Together, these enterprises have generated a vast array of data. In the bulk of these data anorexia is defined objectively by measuring the inhibition of the weight of food consumed by experimental animals in a brief time interval. Consequently, when brought under laboratory control the unit of anorexia is given a simple and unambiguous identity. The simplicity of this notion may be contrasted with the complex interaction of processes which actually regulate the feeding activities of organisms living in natural environments. These processes control the onset, maintenance and termination of eating episodes, the selection of particular nutrients and food items, the expression of preferences and aversions, the long-term temporal patterns of feeding, and the relationship between feeding and other important behaviours. In addition processes controlling the act of eating are usually linked to mechanisms involved in searching or foraging for food. In humans the act of eating is embraced by economic variables and social influences mediated by beliefs, cognitions, and attributions. The act of ingestion can therefore be regarded as a single unit within a much broader system (Blundell 1980). In the laboratory this act of ingestion is normally extracted from the system and examined in isolation. Laboratory studies of food intake therefore make use of a type of *in vitro* representation of behaviour. Accordingly, in considering the clinical applications of psychopharmacological theory of anorectic agents it is worth viewing the relative simplicity of laboratory procedures against the complex system of inter-relating processes within which antiobesity drugs are administered and consumed.

2. CHEMICAL INHIBITION OF FOOD INTAKE

Literature on the pharmacology of feeding indicates that hundreds of chemical agents have been shown to adjust food consumption in laboratory studies. The most frequent form of adjustment is suppression and a varied and extensive range of agents have been reported to reduce food intake when injected, usually by a peripheral route, into experimental animals. These agents include neurotransmitters, hormones, amino acids and other precursors of transmitters, peptides and other neuromodulators, products of digestion, metabolites of enzyme action, certain chemical fractions of blood and urine, and many synthetic compounds such as receptor blockers, reuptake blockers, and direct receptor agonists. Table 4.1 provides some examples of compounds which can reduce consumption. Two questions may be asked about this list of compounds. First, do they suppress food intake by a non-specific disruptive effect on physiology and behaviour, or by a rational intervention into the natural system which serves to match an organism's food intake to its nutritional reuqirement? Second, which sites within the body are the targets for this chemical inhibition?

3. SITES OF ACTION

It is clear that the system which controls food consumption involves a complex network in which metabolic, neural, and hormonal signals are integrated into a coherent pattern. The brain itself can be considered as one system which exists within a much broader physiological system, this in turn is encapsulated in the physical and social environment in which the organism exists (Blundell 1981; Blundell and Rogers 1978). Shifts in food consumption could arise from chemical manipulations at many points in this system. For example, eating may be altered by drugs which act directly on the brain, on gastrointestinal functions (mainly digestion and absorption), on fat metabolism, or on energy expenditure. Many of the agents listed in Table 4.1 appear to have no obvious link with the mechanisms controlling food intake but they affect the system so as to produce changes in consumption. A number of reviewers have drawn attention to the various physiological sites in the body where pharmacological agents may act (Blundell and Rogers 1978; Blundell and Burridge 1979; Peters, Besseghir, Kaserman, and Peters-Haefeli 1979; Sullivan and Comai 1978; Van Itallie, Gale, and Kissileff 1978). Morover, there is currently a great deal of interest in compounds which reduce food intake through a primary peripheral site of action (Sullivan, Comai, and Triscari 1981; Sullivan, Nauss-Karol, and Cheng 1983). Consequently, it cannot be assumed that all chemicals which suppress food intake do so by intervening, physiologically or non-physiologically, with the brain mechanisms responsible for the control of food ingestion.

4. DRUGS WITH CENTRAL ACTION

Among the list of chemicals in Table 4.1 there exists a family of compounds which are usually referred to as appetite suppressants. They are used clinically and are generally believed to have a primary effect on the brain. All of the anorectic drugs clinically available in Europe and America are of this type. Figure 4.1 shows the chemical structures of this family of drugs and it is obvious that most (but not all) anorectic drugs display a superficial resemblance in their

TABLE 4.1 *Various agents which reduce food intake*

Substance	Comments	Authors
'Satiated' blood	Blood from free-feeding rats. 26 ml, injected in 2 ml portions, reduces intake of sweetened milk (30 min test) of rats fed 30 min/d for 24 d.	Davis *et al.* (1967)
'Satietin'	Peptide extracted from human plasma. Injected i.v. or intraventric reduces food intake of rats deprived for 96 h.	Knoll (1980)
'Anorexic urine'	Peptide (pGlu-His-Gly OH) extracted from urine of anorexia nervosa patients. Injected into female free-feeding mice. 50 per cent decrease in daily food intake. Effect persists for year.	Trygstad *et al.* (1978)
Glucose and glycerol	40 per cent glucose by gastric intubation decreases food intake at 2, 7, and 24 h. 40 per cent glycerol decreases food intake over 24 h.	Glick (1980)
(-)-hydroxycitrate	Female rats, fed 3 h/d Lydmycitrate (0.33 nmols/kg) fed orally twice daily. Decreased food intake accounts for reduction in body weight and body lipids.	Sullivan *et al.* (1974)
Threochlorocitrate	Slows gastric emptying. Oral administration reduces food intake in rats and dogs.	Sullivan *et al.* (1981*a*)
Atropine	Rats, 17 h deprived. Sham feeding with gastric fistula (liquid diet). Atropine methyl nitrate (2–250 μg/kg) produces dose-related decrease in food intake.	Lorenz *et al.* (1978)
Adrenaline/Noradrenaline	Rats, high carbohydrate diet for 1 h/d. Adrenaline (0.1, 0.15, and 0.2 mg/kg) and NA (0.1, 0.15 mg/kg) 5 min before. 0.15 dose reduces intake by 71 per cent and 34 per cent respectively.	Russek *et al.* (1967)
Amphetamine/Mazindol	Rats, fed 6 h/d. Drugs given immediately before 2 h test. Amphetamine: 1.25 mg/kg and Mazindol: 7.5 mg/kg. Effects blocked by ventral noradrenergic bundle lesions.	

continued

Table 4.1 continued from page 73

Mazindol, lisuride, Piribidel, nomifensine, Apomorphine	Rats, fed 4 h/d. Drugs injected i.p. at various times before start of 1 h test. All drugs reduce food intake. Effect blocked by pimozide.	Carruba *et al.* (1980)
Serotonin (5-HT)	Zucker fat and lean rats and VMH-lesioned rats, fed 14 g/5h/d. 5-HT (12.5 mg/100g/5 min prior to 2 h test. Decreased eating in all groups.	Bray and York (1972)
Tryptophan	Rats, fed Noyes pellets by eatometer. L-tryptophan (50mg/kg) i.p. Free-feeding rats: decrease in meal size and 24 h intake. 16 h deprived rats: decrease in size of first large meal and increase in post-meal interval.	Latham and Blundell (1979)
5-Hydroxytryptophan (5-HTP)	Rats, 18 h. deprived. 5-HTP (30/60/90 mg/kg) i.p. Dose-related decrease Free-feeding with eatometer, 30 mg/kg 5-HTP decreases meal size and 24 h food intake.	Blundell and Latham (1980)
Fenfluramine, quipazine, ORG 6582, M–CPP (5-HT agonists and uptake blockers)	Rats, fed 4 h/d. Injections immediately before 1 h test. All agents reduce intake.	Samanin *et al.* (1979; 1980) Garattini (1978)
Estrogens	Ovariectomized female rats, unrestricted access to food. Estradiol benzoate –(2.0 μg) and estrone benzoate –(20.0 μg) reduce daily intake.	Wade (1975)
Cholecystokinin	Rats, 5.5 h deprivation. CCK –(2.5 to 40 Ivy dog units/kg) 15 min. before test. Decreased food intake in first 30 min.	Gibbs *et al.* (1973)
Bombesin	Rats, 15 h deprivation, liquid food sham feeding. Synthetic bombesin (2–256 μg/kg) decreases intake in first 15 min of 60 min test.	Martin and Gibbs (1980)

Enterogastrone	Mice, 17 h deprivation, liquid diet. Enterogastrone given 10–15 min after onset of feeding (0.1–1.0 mg) decreases intake for 30-60 min.	Schally *et al.* (1967)
Somatostatin (SS)	Rats deprived early in light part of cycle. 4–5 h later given i.p. 1 μg SS. Decreases intake in 30 min test.	Lotter *et al.* (1981)
Thyrotropin releasing hormone (TRH)	Rats, 20 h deprivation, 1 h test, 30 min data given. TRH(10–20 μg/kg i.p. Injection immediately before test reduces intake.	Vogel *et al.* (1979)
Calcitonin	Rats, 25–50 U/kg decrease 24 h food intake. Rats given 30 min period of eating/day – maximum inhibition when 12.5 U/kg given 4.5/8.3 h before.	Freed *et al.* (1979)
Prostaglandins	$F_{2\alpha}$ – 1 mg/kg i.p. to rats fed for 2 h/d. Suppressed intake for 30 min. Also effective in satiated and partially satiated rats.	Doggett and Jawaharlal (1977*a*).
Prostaglandin precursors	Rats, 22 h deprived. Arachadonic, linolenic and linoleic acids decrease intake for 30–60 min. Effect blocked by indomethacin and paracetamol.	Doggett and Jawaharlal (1977*b*)
Cocaine and coca extracts	Rats, fed ground chow 1 h/d. Cocaine (3.45–27.6 mg/kg) i.p. or p.o. reduces intake. Effect also of chloroform extraction layer of coca.	Bedford *et al.* (1980)
	Rats, 5 h feeding period. Cocaine (10/15/15 mg/kg) i.p. Dose-dependent decrease in first hours. No effect on total intake.	Balopole *et al.* (1979)
Muscimol	Rats, fed condensed milk for 30 min each day. Muscimol (0.5/1.0/2.0 mg/kg) Dose related decrease.	Cooper *et al.* (1980)
Naloxone	Rats, 48 h deprivation, 2 h measurement. Food intake reduced by naloxone 1.0–10.0 mg/kg. No effect in mouse (24 h deprived)	Holtzman (1974)
Tetrahydrocannabinol (THC)	Male rats, fed 6 h/d. THC (2.5 and 5.0 mg/kg) markedly decreases food intake in first 2 h with carry over to nest 4 h.	Sofia and Barry (1974)

chemical structure to the configuration of sympathomimetic amines and can be considered as elaborations of a basic β-phenylethylamine nucleus. This nucleus comprises three basic units: a phenyl ring, a side chain, and a terminal amino (NH_2) group (Fig. 4.1). Consequently, additions to the phenylethylamine nucleus can be made in any of these three units separately, or can take place so as to alter the relationships between the units. This can be illustrated by considering amphetamine which is formed by simply adding a methyl (CH_3) group to the side chain, while the formation of phenmetrazine involves a complex reorganization of the side chain as a whole. The important feature arising from an analysis of the physico-chemical structure of this group of compounds rests on the principle that subtle alterations to the physical structure of a drug may lead to striking changes in the physiological or behavioural effects induced by the drugs. (Biel 1970). For example, amphetamine itself is characterized by a number of well documented effects including, *inter alia*, anorectic action, central-stimulant properties, cardiovascular changes, and a selective effect on certain

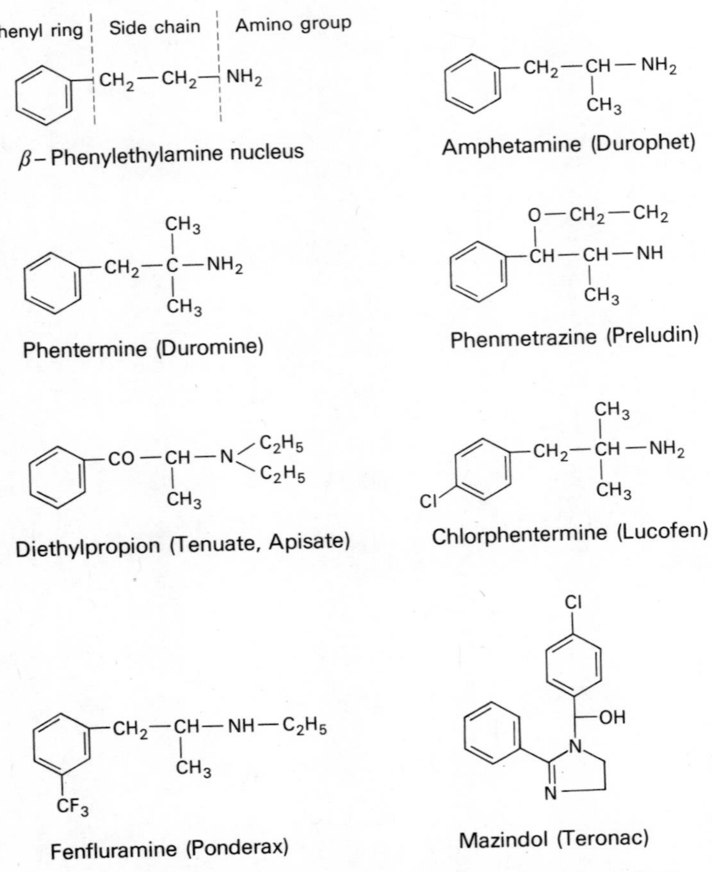

FIG. 4.1 Chemical structures and brand names of anorexic drugs.

neural transmitter agents (particularly the catecholamines). It is now generally considered that, when amphetamine is adminstered clinically, this constellation of effects leads to anorexia being associated with a number of unacceptable bodily and psychological changes which may give rise to the development of dependence upon the drug.

However, it is noticeable that specific chemical changes to the ring, side chain, or amino structure of amphetamine may lead to profound modification of the properties of the drug. In particular, introducing a halogen group into the ring (as in chlorphentermine or fenfluramine), or forming a cyclical structure in the side chain (as in phenmetrazine) leads to marked diminution in central stimulation and cardiovascular activation, with little effect upon anorexic potency. In addition, since the drug may now exert a quite different action upon neurotransmitter agents, the overall profile of the drug has been radically reshaped. Consequently, it is often misleading to compare the effectiveness of drugs by considering their gross structural similarities. For example, despite the apparent superficial structural resemblance between fenfluramine and amphetamine, fenfluramine appears to have fewer properties in common with amphetamine than does mazindol which is structurally quite different.

Accordingly, criteria related to function (what drugs do) rather than to structure (what they do it with) should be used to assess the efficacy of anorectic drugs. Ultimately however, function must be related to mechanism (how they do it), and a confident belief in the selective action of anorectic drugs will only come about with the elucidation of their physiological intervention in the mechanisms underlying the regulation of food intake.

5. NEUROTRANSMITTERS AND ANORECTIC ACTION

Among researchers there is a strong opinion that the action of all the anorectic drugs shown in Fig. 4.1 are mediated by an interaction with monoamine neurotransmitters in the brain. First, most of the anorectic drugs which have been used clinically such as amphetamine, diethylpropion, mazindol, and fenfluramine produce dramatic changes in brain neurotransmitters – particularly the monoamines. Second, pharmacological and neurological manipulations which alter the activity of amines significantly adjust the potency of anorectic drugs to suppress food intake. Third, different neurochemicals appear to be involved in the action of different drugs. Overall, it appears that catecholamines play a dominant role in the actions of amphetamine, phentermine, mazindol and diethylpropion, while the action of fenfluramine appears to be mediated by serotonergic systems (Garattini and Samanin 1976). Consequently, the family of clinically-used anorectic drugs may be divided into two sub-groups.

Most research has been carried out on amphetamine and fenfluramine and, although these drugs are structurally related, they exhibit quite different properties in almost all experimental and clinical situations (Blundell, Latham, McArthur, Moniz, and Rogers 1979). Even their common capacity to diminish food consumption appears to be brought about by different neurochemical mechanisms and through distinctive patterns of behaviour. For these reasons it has been proposed that these compounds can be useful as a drug-pair in exploring the mechanisms underlying the control of food intake. (Blundell and Rogers 1980*a*; Hoebel 1978; Garattini and Samanin 1978).

The differences between amphetamine and fenfluramine can be demonstrated after pharmacological pre-treatment, neurochemical interventions, brain stimulation and lesioning, and by behavioural analysis. In particular, the action of amphetamine is countered by lesions of the ventral noradrenergic bundle (Ahlskog 1974; Quattrone, Bendotti, Recchia, and Samanin 1977), by disruption of catecholamine synthesis using α-methyl-para-tyrosine (Baez 1974) or by dopamine receptor blockers such as pimozide, haloperidol and α-flupenthixol (Blundell and Latham 1980; Garattini and Samanin 1976). Accordingly both noradrenaline and dopamine appear to be involved in the anorexia which follows amphetamine administration. It has been argued that dopamine is more important for amphetamine-induced motor activity than feeding because dopamine antagonists completely prevent the effect of amphetamine on motor behaviour, whereas they only partially counteract the depression of feeding (Samanin, Bendotti, Bernasconi, and Pataccini 1978; Samanin and Garattini 1982). Moreover, it is noticeable that only the effect of higher doses of amphetamine on feeding are attenuated by the relatively specific dopamine-blocking drug pimozide; the anorectic effect of lower doses is unaffected (Burridge and Blundell 1979). From these data a model of amphetamine anorexia has been derived (Fig. 4.2) which assigns roles to both noradrenaline and dopamine in the mediation of amphetamine anorexia - the relative contribution of each neurotransmitter depending on the dose of amphetamine administered (Blundell and Burridge 1979). However the effect of amphetamine on food intake may be a good deal more complicated than this (see Section 00).

In contrast to amphetamine, the action of fenfluramine appears to depend almost entirely on its interaction with serotoninergic mechanisms. Fenfluramine anorexia is countered by pharmacological blockade of serotonin receptors by drugs such as methysergide (Jespersen and Scheel-Kruger 1973; Blundell, Latham, and Leshem 1973) or methergoline (Blundell and Latham 1980; Garattini and Samanin 1976). And the effect is blocked totally (Samanin, Ghezzi, Valzelli, and Garattini 1972) or partially (Clineschmidt, McGuffin, and Werner 1974; Fuxe, Farnebo, Hamberger, and Ogren 1975) following the destruction of central serotonin neurones either by raphe lesions or by the administration of 5, 7-di-hydroxy-tryptamine. However, it should be noted that this latter effect is not always found (Carey 1976; Sugre, Goodlet, and McIndewar 1975). Furthermore, drugs such as chlorimipramine, which inhibit the serotonin-uptake mechanism, have been found to antagonise the anorectic action of fenfluramine (Garattini and Samanin 1976). Although most of these data suggest that fenfluramine exerts its action by the release of serotonin, together with the blockade of re-uptake into the neurone, the actual mode of action operating under particular circumstances may be more difficult to determine. First, fenfluramine is usually administered as the racemate and the two isomers differ in the potency of their effects upon serotonin metabolism (Garattini, Caccia, Mennini, Samanin, Consolo, and Ladinsky 1979). Second each isomer is

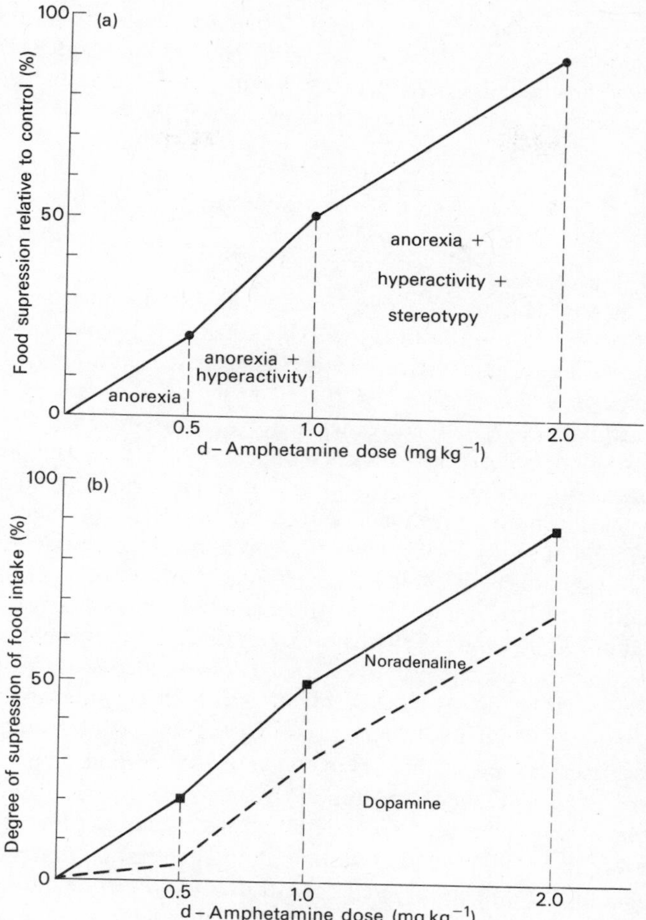

FIG. 4.2 Summary model to demonstrate the dose-related effects of amphetamine. Graph (a) shows the changes in observed behaviour with increasing doses of amphetamine and illustrates how higher doses producing behavioural disturbances could contribute to the inhibition of food consumption and (b) shows the postulated contribution of noradrenergic and dopaminergic systems to the observed behavioural effects.

metabolized to nor-fenfluramine which appears to release serotonin from a different pool (Borsini, Bendotti, Aleotti, Samanin, and Garattini 1982). Consequently, when the parent racemic compound is administered the final action is brought about by four separate pharmacological components. Although all are acting on serotonin mechanisms the overall effect may be complex.

The contrast between amphetamine and fenfluramine illustrates clearly that pharmacologically-induced anorexia may be brought about by involvement of a drug with either catecholamine or serotonin systems in the brain. Among the other commonly-used anorectic compounds, the action of diethylpropion appears to be mediated predominantly by noradrenaline (Borsini, Bendotti,

Carli, Poggesi, and Samanin 1979), whilst dopamine appears to be responsible for the effect of mazindol (Carruba, Ricciardi, Muller, and Mantegazza 1980), although noradrenaline may also be partially involved in mazindol anorexia (Samanin, Bendotti, Bernasconi, Borroni, and Garattini 1977).

6. SUFFICIENT AND NECESSARY CONDITIONS FOR ANOREXIA

The above section demonstrates that noradrenaline, dopamine, and serotonin mechanisms can be involved in drug-induced anorexia. Other evidence confirms this supposition. For example serotoninergic agonists such as quipazine, MK-212 and metachlorophenylpiperazine (MCPP) all bring about a reduction in food intake (Garattini 1978). These and other findings help to establish some role for serotonin in the control of food intake (Blundell 1977; 1979). The involvement of dopamine is supported by the finding that dopaminergic agonists such as *L*-dopa (Sanghvi, Singer, Friedman, and Gershon 1975) piribidel (Carruba, Ricciardi, and Mantegazza 1980), and the ergot derivatives lisuride and bromocriptine (Carruba, Ricciardi, Muller, and Mantegazza 1980) all bring about a strong reduction in food intake in deprived animals. Finally, a role for noradrenaline is implied by evidence that anorexia occurs after peripheral administration of the β-adrenergic agonist salbutamol (Borsini, Bendotti, Thurlby, and Samanin 1982). This effect is blocked by intra-ventricular administration of the β-receptor agonist propranolol. Thus, anorexias produced by amphetamine, fenfluramine, lisuride, and salbutamol may be regarded as models for the further investigation of drug-induced inhibition of feeding; distinct and separate mechanisms appear to underly the anorectic actions of each of these compounds.

Taken together, these findings illustrate only that activation of noradrenergic, serotoninergic or dopaminergic systems constitutes a *sufficient* condition for the inhibition of food intake in experimental animals under specific laboratory conditions. It is also clear, from Table 4.1, that many other chemical compounds also provide sufficient conditions for the occurrence of anorexia. Most notable among these are the actions of gamma-amino-butyric acid (GABA) agonists and opiate antagonists. So far, the discovery of a common component in the above drug-induced anorexias, which would constitute a *necessary* conditon, has not yet been made. It is possible that various inputs from the physiological system governing eating (e.g. level of blood glucose or fatty acids, feedback to brain from release of gut peptides, peripheral information mediated by vagal afferents) are each mediated in the brain by different populations of receptors and neurotransmitter systems. Thus the occurrence of different types of drug-induced anorexia can be understood (Blundell 1981). It should be remembered however that in some instances at least drug-induced inhibition of eating may be related to the particular physiological and laboratory conditions operating. A wider range of testing situations should reveal the robustness of neurotransmitter models and may reveal the necessary conditions required for the occurrence of anorexia under *all* circumstances (see Section 00).

7. DRUGS AND THE CENTRAL MECHANISMS FOR THE CONTROL OF FOOD INTAKE

The research discussed in the preceding section has revealed that pharmacological or neurological interference with specific neurotransmitter systems can antagonize the anorexia induced by specific drugs. What is the appropriate interpretation of these findings? The philosophy usually adopted for understanding these effects exists in a strong and weak form. One interpretation (the *stronger*) is that the neurochemical systems manipulated to antagonize drug-induced anorexia must be implicated in the neural control of food intake. Another (the *weaker*) is that the particular neurochemical adjusted merely mediates the effect of the drug. Adoption of the stronger interpretation must overcome the logical objection that, before a neurochemical mechanism can be implicated in feeding control, it must be demonstrated that the drug-induced inhibition of intake makes use of the natural feeding control system. (Blundell and Latham 1982). The weaker interpretation is less controversial; it does not depend on the drug operating through normal channels. One way to establish the stronger form would be to demonstrate that centrally-acting anorectic drugs can act via the brain mechanisms believed to control food intake. What is the evidence for this?

For more than a quarter of a century the focus for brain mechanisms controlling feeding has been the hypothalamus. Initially, the action of anorectic drugs was explained in terms of a mediation by the ventro-medial or lateral zones of the hypothalamus (Blundell and Latham 1979). However, the dual-centre theory of feeding has been considerably revised by the advent of technological advances and new findings (Blundell 1982). The most sophisticated elaboration has been proposed by Hoebel and Leibowitz (1981) based on studies using topical micro-injections of chemicals. This work has identified two critical zones - the perifornical area of the lateral hypothalamus and the more medial paraventriclar nucleus - which respond quite differently to micro-injections of neurotransmitters and other drugs. The current view is set out in Fig. 4.3. In summary, injections of agents which facilitate α-adrenergic activity induce feeding when injected in the paraventricular area. (Leibowitz 1978; Leibowitz, Arcomano, and Hammer 1978). On the other hand, agonists of dopamine and β-adrenergic receptors depress feeding when injected in the perifornical zone (Leibowitz and Rossakis 1978; 1979). These local effects have been shown to depend upon the integrity of the dorsal and ventral portions of the central tegmental tract Leibowitz and Brown 1980*a,b*). It may be inferred that these critical zones in the hypothalamus, together with their neuronal connections, constitute the neural substrate for the expression of feeding activity.

Of particular importance here are the effects of micro-injections of anorectic drugs. Injections of amphetamine or mazindol into the perifornical area suppress food intake, an effect also shown by dopamine, epinephrine and isoproterenol (Leibowitz and Rossakis 1979). In the paraventricular neucleus injections of serotonin or 5-hydroxytryptophan inhibit eating (Leibowitz and Papadakos 1978) as does fenfluramine (S. Leibowitz, personal communication). Accordingly, these data demonstrate the presence of active sites in the hypothalmus

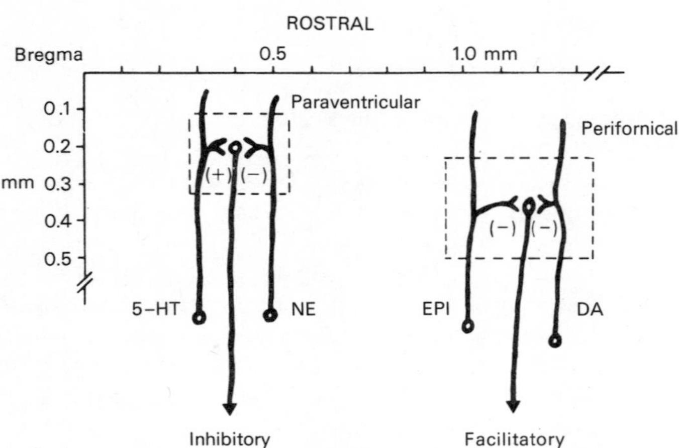

FIG. 4.3 Summary of hypothalamic neurotransmitter pathways involved in the control of food intake (Hoebel and Leibowitz 1981).

for anorectic drugs injected locally and it follows that these same sites *could* provide the critical zones of action for anorectic drugs administered systemically. This basic model of hypothalamic neurones has been modified to include new data on the micro-injection of endorphins and other opioids (Leibowitz and Hor 1980; Hoebel 1984). Indeed, a central model incorporating sites of action for many amines, peptides and other agents has been proposed by Morley (1980) (Morley and Levine 1983), whilst a temporal rather than spatial model has been proposed by Blundell (1981, 1982). Taken together, these conceptualizations provide a summary of the evidence for the action of anorectic agents via brain sytems controlling food intake.

Very recently, a further possible central mechanism of action for anorectic agents has captured attention. Using *in vitro* methodology, saturable and stereo-specific binding sites for radioactively labelled amphetamine were located in the hypothalamus. Other phenylethylamine derivatives also bind to synaptosomal membranes from this hypothalamic region (Paul, Hulihan–Giblin, and Skolnick 1982). Indeed, several amphetamine derivatives including *p*-chloro-amphetamine and fenfluramine were more effective than amphetamine in displacing (+)-[^{3}H] amphetamine binding. More interestingly, a comparison was made between the affinity of phenylethylamine derivatives for the (+)-[^{3}H] amphetamine-binding site with the ED_{50} values for anorectic potency taken from another laboratory (Cox and Maickel 1972). The authors reported a correlation of 0.97. At the present time these findings have not yet been confirmed, and further experiments to demonstrate the physiological significance of the specific amphetamine binding sites have not yet been reported. However, if these results are replicated, they constitute evidence for the presence in the brain of specific receptors apparently mediating anorexia, and are quite unrelated to other behavioural functions such as motor stimulation. For the moment the claim for specific 'anorectic receptors' must be treated cautiously. Indeed, in view of the weight of evidence indicating quite separate mechanisms of action for drugs such as amphetamine and fenfluramine, it does seem unlikely that they should both act via the same sub-population of 'anorectic' receptors.

8. ISSUES IN ANIMAL RESEARCH

Evidence about the effectiveness and mechanism of action of anorectic agents is invariably based, initially, on research on animals - usually the laboratory rat. However, there exist a number of methodological problems which hinder the interpretation of the animal data, particularly when applying these to clinical conditions in humans. Certain of these problems are briefly discussed below.

A. Food deprivation

Many studies on drug-induced anorexia employ a simple experimental strategy in which a rat deprived of food for 16, 24, or 48 h is allowed access to food for a brief interval (usually 1 or 2 h). The depression of the amount of food consumed during this interval is regarded as a measure of the anorectic potency of the drug. It has been argued on many occasions that this procedure may present a barrier for further understanding of the effects of drugs on feeding (Blundell 1981; Blundell and Latham 1978; 1979*b*; 1982). In particular, severe food deprivation constitutes a powerful physiological trauma which may create circumstances for the appearance of abnormal behaviour, and modify drug action in an unknown way by altering brain chemistry. Consequently, drugs administered to deprived animals may be intervening in atypical brain metabolism. This problem can be further confounded when, for example, deprived animals are used in tests for anorexia, whilst satiated rats are used for neurochemical analyses. Indeed, it has been demonstrated that food deprivation produces a depletion of hypothalamic norepinephrine, and antagonises the effect of amphetamine on this substrate (Glick, Walters, and Milloy 1973). Since food deprivation itself affects mechanisms involved in feeding, it may be expected that different results will be obtained - behaviourally and neurochemically - from drug tests on deprived and non-deprived animals.

B. The dependent variable - weight of food consumed or behavioural profile

Another limiting factor is the tendency to use weight of food consumed (in a brief interval of time) as the measure of a drug's effectiveness. Such a measure is unreliable since it offers no information about why the animals failed to eat a greater amount. Obviously, rats may fail to eat for a wide variety of reasons, including excessive changes in motor activity (sedation or excitement), the occurrence of abnormal behaviour (e.g. stereotypy, backward walking), disruption of the natural sequencing of behavioural units, in addition to the intended effect of a change in the strength of hunger or satiation. Consequently, a simple measure of the weight of food which has disappeared from a dish or hopper in a discrete interval of time may conceal information about the manner in which a drug has adjusted consumption. One way to check the disruptive effect of drug administration on behaviour, and to throw light on mechanisms of action, is to examine the actual changes in the structure of behaviour which occurs whilst an animal has access to food. This may be carried out in two ways. First, to use video-recording or direct observation of behaviour to reveal changes in the *micro-structure* of behaviour. In the rat feeding is a discontinuous activity, and even during the course of a concentrated period of feeding animals will spend short amounts of time eating, grooming, sniffing, walking about, and possibly drinking. These small bursts of specific activities constitute a sequence of behaviour and the frequency and duration of these units can be recorded and measured (Wiepkema 1971*a,b*). When this procedure is carried out following the administration of anorectic drugs, a detailed record can be made of the effects upon such feeding parameters as total food intake (g), duration of time spent eating (min), number of eating bouts (*n*), size of bouts (g), duration of bouts (min), together with the overall and local rates of eating (g/min) (Blundell and Latham 1977; 1978; 1979*b*). In addition measures can be made of non-feeding activities such as grooming, locomotor activity and sedation. Such recordings can be used to detect unusual or disruptive intrusions into the behavioural sequence, and to indicate how the fine structure of feeding is adjusted by the anorectic agents (Blundell and Latham 1978, 1979*b*). In turn, these adjustments in behavioural structure can help to reveal the mechansism by which the anorectic drug reduces the weight of food consumed.

An alternative technique involves the recording and measurement of long-term feeding patterns in *free-feeding* rats never subjected to periods of food deprivation. Such continuous monitoring leads to the identification of meals from which may be computed meal size, meal duration, meal frequency, inter-meal rates of eating, and changes in structure. (Blundell and Latham 1982; Blundell 1982). The procedure appears to have been used in drug studies about 17 years ago (Borbely and Waser 1966) when the active drug (amphetamine) was delivered via the rat's drinking water. A more recent study, which compared the effects of amphetamine and fenfluramine, demonstrated that these drugs displayed quite different profiles when meals were monitored continuously over 24-h periods in non-deprived rats (Blundell and Leshem 1975; Blundell, Latham, and Lesham 1976). Interestingly, the effect of fenfluramine was characterized by a reduction in meal size which suggested that the drug was acting to promote the process of satiation and thereby to cause an early termination of eating. This effect has now been confirmed many times (Blundell and Latham 1978; Davies 1976; Grinker, Drewnowski, Enns and Kissileff 1980; Burton, Cooper, and Popplewell 1981; Davies, Rossi, Panksepp, Bean, and Zolovick 1983). This type of analysis provides qualitative information about feeding behaviour over long periods of time and permits alterations in behavioural parameters to be matched against the time course of drug concentration in blood (Blundell 1982; Blundell, Campbell, Lesham, and Tozer 1975). Moreover, the real power of this technique is to provide clues about how the drug may influence the processes of hunger, appetite and satiation (Blundell 1979) which underly the organization of feeding behaviour and the control of intake.

C. Novelty, variety, and palatability of food

Traditionally, pharmacological studies on feeding have used laboratory rats allowed to consume only the bland and balanced diet of laboratory chow. However, rats are omnivorous animals which exploit many different food sources through foraging and hoarding. Accordingly, it is likely that under natural circumstances rats encounter a number of different types of food differing in taste, texture and nutrient density. The tendency of rats to eat more of a tasty and varied diet has been experimentally exploited to produce a form of dietary-induced obesity (Sclafani and Springer 1976; Sclafani 1978). Because of the use of novel and varied food items in anorectic drug research it will now be possible to (a) distinguish between drug effects on hunger and appetite (Blundell 1979; Blundell and Rogers 1980*a,b*) and (b) assess the action of drugs on the hedonic value of food (Blundell 1982). It has also become clear that not all anorectic drugs suppress intake equally when offered different types of food. The use of varied diets in drug research is essential to investigate the processes underlying the action of anorectic drugs, and to increase confidence in generalizing from laboratory to clinical studies.

D. Dietary-induced obesity

The use of varied and palatable diets has encouraged interest in a form of experimental obesity which appears to have features in common with the development of obesity in man. Accordingly, the effectiveness of anorectic drugs to inhibit intake of highly palatable food, and to retard or suppress the occurrence of dietary induced obesity, becomes clinically relevant. It has been demonstrated that the opiate receptor blocker naloxone exerts a greater suppressive effect on the consumption of a snack-food diet than of laboratory chow, at least during the dark (Apfelbaum and Mandenoff 1981). This effect also occurs with naltrexone (Mandenoff, Fumeron, Apfelbaum, and Margules 1982), and appears to indicate a more potent anorectic effect when the diet induces hyperphagia. However, this effect is not unique to opiate blockers. Interestingly, amphetamine is less potent with a varied diet than with laboratory chow but the action of naloxone is matched by fenfluramine (Bowden, White, and Tutwiler 1983). Moreover, fenfluramine has been shown to suppress the level of dietary-induced obesity (Kirby, Pleece, and Redfern 1978) and the *d*-isomer is more effective in the plateau than the dynamic stage of obesity development (Hill, Rogers, and Blundell 1983). In other words *d*-fenfluramine is more effective at counteracting obesity once it has developed than it is in retarding its onset. Now that both fenfluramine and naloxone have been shown to exert greater anorectic potency with highly palatable food. wheras amphetamine and (-) hydroxycitrate do not, it becomes important to examine the effect of all anorectic agents in this model of experimental obesity. Effectiveness displayed here may be predictive of clinical efficiency.

E. Total calories versus nutrient selection

A considerable body of evidence indicates that omnivorous animals, feeding from a variety of food sources, possess the ability to select and to qualitatively monitor their intake of certain nutrients (Rozin 1976; Overmann 1976). For example, rats allowed to select from a cafeteria array of separate dietary components such as protein, fat, carbohydrate, vitamins, and minerals are able to maintain a balanced intake of essential elements (Richter 1943). In addition, this dietary self-selection can be influenced by the ambient temperature (Leshner, Collier, and Squibb 1971), changing hormonal states (Leshner, Siegel, and Collier 1972) activity level (Collier, Leshner, and Squibb 1969) and the availability of water (Overmann and Yang 1973; Corey, Walton, and Weiner 1978). Indeed, it appears that dietary self-selection is a fundamental characteristic of feeding behaviour in animals which becomes more clearly apparent when functional demands are placed on the system.

Until recently, this phenomenon had been ignored in pharmacological investigations of feeding in which animals are generally maintained on a single composite diet containing a balanced mixture of essential nutrients. However, interest in pharmacological aspects of voluntary self-selection has been promoted by theoretical developments concerning the role of neurotransmitter systems in the regulation of protein and carbohydrate intake. It has been proposed that the concentration of the transmitter serotonin in the brain is dependent upon the ratio of tryptophan to neutral amino acids in the plasma (Fernstrom and Wurtman 1973). The nature of the diet exerts a major influence over this plasma ratio, and it has been demonstrated in rats that a high carbohydrate meal can lead to increases in brain tryptophan and serotonin (Fernstrom and Wurtman 1972). It has therefore been proposed that serotonin-containing brain neurones may function as 'ratio-sensors' – the rate of neurotransmitter synthesis in these neurones varying with the nutrient composition of the diet. Although there exists some opposition to this idea (Neckers, Biggio, Moja, and Meek 1977), it has been argued that serotonin neurones could discriminate between the metabolic effects of various diets (Fernstrom and Wurtman 1973). One implication of this process is that the feedback effect from neurotransmitter synthesis to feeding behaviour may involve qualitative rather than quantitative adjustments in food intake. This means that neurotransmitter activity will influence an animal's choice of nutrients. In keeping with this idea Ashley and Anderson (1975) have demonstrated that in the weanling rat the ratio of tryptophan to neutral amino acids in the plasma is related to the amount of protein self-selected by the rat. In turn this has led to the suggestion that serotonin neurones participate in feeding, not by regulating total caloric intake, but by adjusting protein (Anderson 1977; 1979), or by controlling the balance of protein and carbohydrate in the diet (Fernstrom and Wurtman 1974).

This hypothesis lends itself readily to test by pharmacological agents, and allows drugs to be used as tools to further probe the complex structure of feeding. Moreover, experimenters are obliged to allow animals to voluntarily self-select their intake of particular dietary components. This demand has led to the development of an experimental procedure in which animals are faced with two or more food containers holding varying concentrations of the major macro-nutrients protein, carbohydrate, and fat. In a typical two-container design animals are allowed to choose between protein values of 5 and 45 per cent, or 15 and 55 per cent, or 0 and 60 per cent in isocaloric diets. Accordingly, by moderating their eating from a particular source, animals can adjust their elective consumption of protein and carbohydrate. In one of the first pharmacological studies using this paradigm it was demonstrated that different anorectic drugs exerted distinctive effects on the pattern of selection (Wurtman and Wurtman 1977). Fenfluramine and fluoxetine which increase the synaptic activity of serotonin, displayed a protein-sparing effect in weanling rats; that is, they reduced total food intake but actually increased the proportion of protein consumed. In contrast, amphetamine gave rise to an equal suppression of protein and total caloric intake. In an extension of this study using adult animals and allowing the rats to feed freely, fenfluramine maintained, but did not spare, protein intake whilst amphetamine brought about a severe suppression of protein consumption (Blundell and McArthur 1979; McArthur and Blundell 1983). Recently, it has been shown that the depletion of brain serotonin by systemic injections of *p*-chloro-phenlyalanine, intraventricular administration of 5,7-dihydroxytryptamine, or by electro-thermal lesions of the medial raphe nuclei, lead to a selective decrease in protein intake (Ashley, Coscina, and Anderson 1979). In turn, this work is complemented by the finding that drugs which are believed to enhance central serotonergic transmission selectively suppress carbohydrate consumption by rats (Wurtman and Wurtman 1979*a*). Under certain circumstances pharmacological manipulation of serotonin metabolism can adjust the selection of macro-nutrients by rats, although it is currently not clear whether the primary effect is on protein or carbohydrate intake (Wurtman and Wurtman 1979*b*).

More recently, the phenomenon of nutrient selection has been complicated by the finding that clonidine, an alpha-adrenergic agonist enhances food consumption and appears to specificially stimulate protein intake (Mauron, Wurtman, and Wurtman 1980). This effect was interpreted as an action upon presynaptic receptors, and it was suggested that central noradrenergic neurones participate in the mechansims controlling appetites for protein. Accordingly, it seems that both serotonin and noradrenaline systems may be involved in the control of protein–carbohydrate intake. However, this picture is further complicated by the report that injections of clonidine actually increase the preference for carbohydrate following peripheral or central administration (Fahrbach, Tretter, Aravich, McCabe, and Leibowitz 1980). Since similar effects have been obtained after noradrenaline injections into the paraventricular nucleus (Tretter and Leibowitz 1980), these findings tend to implicate central noradrenaline systems in the control of carbohydrate rather than protein consumption.

Using a different design in which pure samples of protein carbohydrate and fat are offered in the selection test, it has been reported that insulin (Kanarek, Marks-Kaufman, and Lipeles 1980), and gonadal hormones (Kanarek and Beck 1980) influence diet selection. Moreover, whilst morphine led to an increase in fat intake (Marks-Kaufman and Kanarek 1980), amphetamine selectively depressed consumption of this commodity (Kanarek, Ho, and Meade 1981). These data suggest that fat, protein and carbohydrate may be under the control of complex neurochemical relationships in the brain.

Accordingly, the dietary self-selection model provides an experimental arena for the deeper investigation of anorectic agents. At the present time it appears that most centrally-acting drugs exert some selective effect on macro-nutrient consumption, in addition to an effect on total caloric intake. Moreover, these selective actions need not necessarily be related to underlying mechanisms of macro-nutrient regulation.

F. Models of overeating

The 'classical' method of testing anorectic agents has invariably used food restriction (absolute fast or cyclic feeding regime) to induce an experimental animal to eat at a time convenient for laboratory investigators. However, there exists a variety of procedures for promoting overeating in the short or long term which could be profitably used for evaluating the action of anorectic drugs. A selection of techniques is set out in Table 4.2. It is obvious that drugs which effectively suppress deprivation-induced eating may not necessarily have an effect on alternative processes underlying the induction of eating. For example, amphetamine which is a potent inhibitor of eating following deprivation does not suppress eating induced by stress (Antelman, Caggiula, Black, and Edwards 1978) or by insulin injections (Carruba, Mantegazza, Muller, and Ricciardi 1981). More extensive use of a range of overeating models will serve two purposes. First, it will test the robustness of the inhibitory action produced by an anorectic drug, and thereby increase confidence that it will be effective in man. Secondly, these procedures will help to throw light upon the mechanisms underlying the action of the drugs. It is important for purposes of screening, and for deeper investigations, that researchers turn away from the convenient, but problematic, technique of food deprivation and take up alternative models.

G. The interpretation of amphetamine anorexia

For more than forty years amphetamine has been regarded as the standard anorectic drug. It is of course true that amphetamine can reduce food intake, but in other domains of research locomotor activity and stereotypy have been considered as the fundamental behavioural parameters (Costa and Garattini 1970). Accordingly, there is no *prima facie* reason for believing that the decrease in food intake brought about by amphetamine is achieved through an interaction with some primitive system which serves to match food consumption to nutritional requirements.

Apart from the powerful effects of amphetamine on the autonomic nervous system which may subserve the transmission of signals relevant for feeding, the action of amphetamine is readily detectable at the behavioural level (Cole and Gay 1974; Cole 1978). For example, using amphetamine as a model for hyper-

TABLE 4.2 *Experimental procedures useful for testing the effects of anorexic agents on overeating by rats*

Techniques for inducing overeating	Drug investigated
Addition of sucrose to diet	Naloxone
Cafeteria foods	Naloxone, naltrexone, amphetamine, fenfluramine
Electrical stimulation of lateral hypothalamus	Amphetamine, fenfluramine
Noradrenaline injections into paraventricular nucleus	Fenfluramine
Ventro-medial hypothalamic lesions	Amphetamine, fenfluramine, chlorphentermine
Hypothalamic knife cuts	— — — —
Muscimol injections into medial raphe nuclei	Amphetamine, fenfluramine
Lesions of ventral noradrenergic bundle	Amphetamine, fenfluramine, di-ethylpropion
Injections of diazepam	Naloxone, amphetamine
Injections of yohimbine	Naloxone
2-Deoxy-d-glucose administration	Naloxone, fenfluramine, amphetamine
Insulin injections	Naloxone, fenfluramine, amphetamine
Stress-induced (tail-pinch) arousal.	Amphetamine, fenfluramine, naloxone

activity in rat, Norton (1973) has demonstrated that amphetamine causes dose-related changes in the frequencies of certain behavioural acts, and also reduces the duration of the acts once initiated. This disruption of normal behavioural sequences is also reflected in the general hypothesis of the 1975 version of the Lyon and Robbins (1975) theory. This hypothesis stipulates that, as the concentration of amphetamine increases in the central nervous system, 'the organism will tend to exhibit increasing response rates within a decreasing number of response categories'. (p.85). Together, these observations and principles provide a rationale for the blockade of feeding through the re-adjustment of the normal pattern of behaviour.

Moreover, in recent years evidence has revealed that amphetamine does not invariably produce a consistent effect on food consumption. For example. in deprived rats amphetamine usually increases the rate of eating whilst decreasing total food intake (Blundell and Latham 1978; Blundell and Latham 1980). More important, amphetamine has been shown to produce paradoxical increases in food consumption. This has been observed when very low doses of the drug are administered to mice (Dobrzanski and Doggett 1976) or rats (Blundell and

Latham 1978) which have not been subjected to the usual deprivation procedures. In addition this phenomenon has been shown to occur in cats (Wolgin, Cytawa, and Teitelbaum 1976) and rats (Stricker and Zigmond 1976) which have been subjected to lateral hypothalamic lesions, and it has occasionally been reported in food deprived animals (Holtzman 1974). The explanation favoured to account for these unexpected results is one involving the use of an intervening construct such as arousal or activation. Accordingly, the effect of amphetamine on feeding will vary according to the internal state of the organism. This idea is similar to that put forward by Eichler and Antelman (1977) to explain how apomorphine could give rise to feeding or anorexia, depending on the nutritional status of the animal. A pressing need is to find the appropriate interpretation for the enhancement and suppression of food intake brought about by amphetamine under differing circumstances. The model in Fig. 4.4 suggests one simple way in which an explanation can be constructed. The diagram shows how different doses of amphetamine could give rise to varied, or even opposite, effects depending on the value of some state of the organism labelled construct 'x'. At the present time it is not necessary to seek a deeper identity for this construct, although since the strongest evidence for contrasting behavioural actions is found with deprivation and hypothalamic damage, it is tempting to suggest that 'x' represents some central arousal state embodied in catecholaminergic systems.

Such an interpretation proposes a non-specific mediation of the effects of amphetamine on feeding in which the action of the drug would be only obliquely related to a nutritional control system. However, it is also possible, as certain reports suggest (see Section 5) that at certain doses amphetamine directly

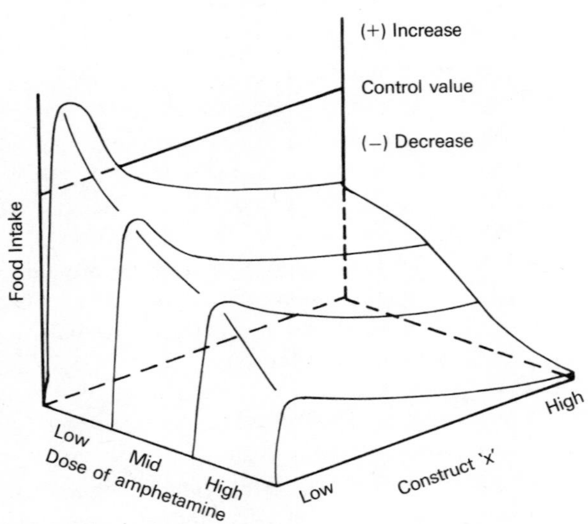

FIG. 4.4 Conceptualization to describe the wide range of effects of amphetamine on food intake.

influences an endogenous feeding substrate. As more data become available, a more complex network of mechansims will be required to describe the effects of amphetamine on feeding. However, it is clear that the action of amphetamine varies with the dose of the drug and with other features of the environmental circumstances. It therefore seems appropriate that the effects of amphetamine on feeding should be interpreted cautiously, and that amphetamine should now cease to be used as the standard reference compound in the pharmacology of anorexia.

H. Central or peripheral action

Although there is clear evidence that the anorectic drugs discussed in this chapter exert strong effects on brain neurotransmitters it remains possible that some part, or even all, of the anorectic effect could be due to an action outside of the brain. After all, the monoamine neurotransmitters exist in abundance in the periphery. For amphetamine it has been shown that coeliac vagotomy (Tordoff, Hopfenbeck, Butcher, and Novin 1981) and truncal vagotomy (Tordoff, Novin and Russek 1980) attenuate the anorectic effect of low doses. It has been argued that a part of the anorectic action of amphetamine is mediated by noradrenergic mechanisms of the sympathetic nervous system. The detection of such effects may depend upon the dose of drug administered.

For fenfluramine it has been argued that the anorectic action, at least in non-deprived animals, may be related to a slowing of gastric emptying (Davies 1976; Davies, Rossi, Panksepp, Bean, and Zolovick 1983). Fenfluramine does delay stomach emptying, and was shown to be most effective in lengthening the post-meal interval when given immediately after feeding. These results suggest that fenfluramine controls feeding through short-term signals related to food in the upper gastrointestinal tract. This effect is consistent with the postulated effect of fenfluramine on satiety (Blundell and Leshem 1975; Blundell, Latham, and Leshem 1976) and with a role for peripheral serotonin in the control of feeding (Blundell 1979*b*).

I. Conclusions

The last 10 years have seen marked developments and shifts in emphasis in research on anti-obesity drugs. Owing to the risk of abuse of amphetamine-like compounds, some research groups have directed attention to drugs with a largely peripheral mode of action (Sullivan, *et al.* 1983). Currently, a good deal of effort is involved in the search for thermogenic compounds (Levin 1983). During this same period developments have also occurred in research on centrally-acting agents. The synthesis of fenfluramine more than 15 years ago radically changed thinking about the pharmacology of anorexia. First, the drug demonstrated that anorexia could be produced without marked CNS stimulation; indeed, if anything, the drug gives rise to sedative effects. Second, the mechanism of action of the drug drew attention to the role of the neurotransmitter serotonin in the control of food intake. Third, the behavioural adjustments brought about by the drug suggested an involvement with the process of satiation rather than hunger.

The last decade has witnessed the rise of the opiates as compounds believed to regulate or modulate feeding via central mechanisms, and during this period central neurotransmitter models of food intake control have advanced thinking beyond the dual-centre theory. Along with these technological and theoretical developments have come certain practical advances. There is now available a much wider range of testing procedures, experimental models and design strategies. Centrally-acting anorectic agents can be subjected to broader and more penetrating examination, and their modes and mechanisms of action more thoroughly investigated. Research on animals will continue to offer new insights for the furtherance of understanding of centrally-acting agents, and for the mechanisms through which they exert their actions. This research is vital for the advancement of knowledge about the possible effects of such drugs in man.

9. HUMAN PSYCHOPHARMACOLOGY OF ANOREXIA

The experimental study of drug-induced anorexia in man has a history stretching back more than 40 years (Silverstone 1982) to the first studies on the effects of amphetamine (Bahnse, Jacobsen, and Thesleff 1938). Compared with animal research, the number of investigations is few and, owing to technical difficulties, the mechanisms of action have been less thoroughly investigated. Indeed, until fairly recently, the action of anorectic drugs in man had to be deduced from research on animals. In recent years theoretical advances derived from animal experiments have generated sophisticated studies on mechanisms and modes of action in man (Silverstone 1982; Silverstone and Kyriakides 1982).

A. Basic techniques

A minimal requirement in the assessment of anorectic drug activity is that some measure should be made of subjective experiences of hunger, appetite or satiation together with an objective record of food consumption (Hill and Blundell 1982/83). Frequently, researchers have been satisfied with a measurement of the subjective sensation of hunger, and this has been translated into objective data through ratings on a visual analogue scale (Silverstone and Stunkard 1968). Usually, the effectiveness of an anorectic drug has been assessed by the capacity of the drug to lower ratings of hunger. In certain cases this has been taken as evidence of the potency of the drug in the absence of measures of food consumption. Visual analogue ratings of hunger may be supplemented by recordings of bodily sensations of hunger or satiety (Monello and Mayer 1967; Garfinkel 1974; Blundell, Latham, McArthur, Moniz, and Rogers 1979) and by ratings of appetite and satiation (Blundell and Rogers 1980). However, it is clearly desirable that caloric intake should be recorded and this can be done directly or indirectly. Indirect techniques involve self-recording procedures and require subjects to maintain an accurate record of the nature, amount, time and place of all food consumed (Beaudoin amd Mayer 1953). From these reports an eating profile can be computed but this procedure is clearly best suited to the long-term monitoring of intake. A more difficult strategy is to monitor eating more directly and for shorter periods – a procedure of greater value in the assessment of short-lived effects of centrally-acting drugs. This can be achieved through a variety of procedures including the video-taping of all eating episodes (Rogers and Blundell 1979), using a strain gauge or load cell to measure the weight of a plate of food continuously (Kissileff, Klingsberg, and Van Itallie, 1980), using eating utensils specially equipped for telemetering (Moon 1979), making continuous measurements of jaw movements (Bellisle and Le Magnen 1980), using reservoirs to pump (Hashim and Van Itallie 1964) or dispense (Jordan, Weiland, Zebley, Stellar and Stunkard 1966; Pudel and Oetting, 1977) liquid food, or using snack-food dispensers specially programmed to record the removal of solid-food items (Silverstone and Fincham 1978; Wurtman and Wurtman 1981). In addition to these techniques the measurement of salivation induced by food stimuli has been developed as an index of appetite or motivation to eat (Wooley, Wooley, and Williams 1976). The choice of technique depends on the required duration of monitoring, and on the fineness of detail (about eating behaviour) that is required. Of course, certain of these techniques do not measure intake 'directly', but they do record behaviour intimately connected with the act of eating. These procedures are sensitive to changes in total food intake produced by a drug like amphetamine (Silverstone and Fincham 1978), phenylpropanolamine (Hoebel 1974), and cyproheptadine (Saleh, Yang, Van Itallie and Hashim 1979), and by hormones such as cholecystokinin (Kissileff, Pi-Sunyer, Thornton, and Smith 1981).

B. Major findings

In the last 40 years approximately 30–40 reports have appeared documenting the effects of centrally-acting anorectic agents on food intake or subjective experiences of hunger and appetite (not including clinical trials in which only body weight has been measured). (Blundell and Rogers, 1978; Silverstone and Kyriakides, 1982).

The starting point was the observation that patients in clinical trials of amphetamine as a treatment for depression and narcolepsy lost weight and complained of diminished appetite (Bahnsen, Jacobsen, and Thesleff 1938; Davidoff and Reifenstein 1937; Jacobsen 1939; Jacobsen and Wortstein 1939; Nathanson 1937). In one study (Jacobsen 1939) 34 per cent of subjects given 15 mg amphetamine reported having 'poor appetite' in questionnaires, whilst in a second study (Nathanson 1937) 50 per cent of subjects receiving 20 mg amphetamine noted decreased appetite in a general questionnaire. Harris, Ivy, and Searle (1947) however, were the first to demonstrate the food-suppressant effect of amphetamine. In seven obese and two non-obese subjects calorie intake was reduced by between 8 per cent and 25 per cent with amphetamine (5--7.5

mg/day). On the other hand, ten normal male subjects on a constant food intake regime (3000 cal/day) given 5–10 mg amphetamine every day, lost on average less than one pound in eight weeks. These authors did not, however, measure the effects of the drug on hunger.

Although some workers subsequently failed to detect any effect of amphetamine on food intake and hunger ratings (Bernstein and Grossman 1956; Kroger 1972) several authors have more recently re-examined the appetite-suppressing effects of amphetamine in humans (Hollister 1971; Silverstone and Stunkard 1968; Rogers and Blundell 1979). Dexamphetamine, compared with placebo, significantly reduced hunger ratings of subjects of normal weight, as measured by a visual analogue scale, within 105–120 min after administration, but there was no difference between 5 mg, 10 mg, 15 mg, or 20 mg doses of the drug (Silverstone and Stunkard 1968). However, there was a weak, but significant, positive correlation between the hunger-reducing effects of the drug and reduction in calorie intake (mean reduction 18 per cent). In another study 15 mg dexamphetamine reduced hunger in mildly food-deprived subjects between 8.00 am and lunch, whereas hunger ratings with placebo increased during this time. Food intake was reduced by 22 per cent (Hollister 1971).

Wooley and Wooley (1973) used stimulus-induced salivation as a measure of appetite; they reported that 10 mg amphetamine suppressed the salivary response to food of normal subjects an hour after drug ingestion, and before any effects on hunger ratings were seen (Wooley *et al.* 1976). Recently, we have confirmed these results showing that calorie intake, hunger ratings, and salivation of non-obese subjects were reduced at three hours by 10 mg amphetamine. Hunger ratings were still significantly lower seven hours after taking the drug compared with placebo. However, we found only low correlations between the change in various subjective ratings of hunger and the change in food intake brought about by amphetamine. Interestingly, amphetamine appeared to affect the 'willingness of our subjects to initiate consumption'. Thus, with this drug subjects were slower to take the first mouthful after having sat at the table to eat, compared with when they received fenfluramine or placebo (Rogers and Blundell 1979). This result is consistent with animal studies showing that latency to eat is increased by amphetamine.

The effects of fenfluramine on human feeding have also been investigated in some detail. One study compared fenfluramine (60 mg/day), phenformin (50 mg/day) diethylpropion (25 mg/day) and phentermine (90 mg/day) in thirteen normal weight male subjects. Over four weeks there was a downward trend in weight on all drugs, but no effect on appetite was seen except a slight decrease with fenfluramine compared to scores with placebo. Immediately on stopping the drugs some subjects reported 'exceptional hunger' (Malcolm, Mace, and Ontar 1972). However, in a clinical trial both fenfluramine (60 mg/day) and phentermine (30 mg/day, Duromine) strongly depressed appetite consistently over 28 days, compared with four weeks on placebo (Steel and Briggs 1972). In an unusual study (Pudel 1975) fenfluramine was given to obese, latent-obese and non-obese subjects. The tablets were clearly labelled as an appetite suppressant or placebo, a quarter of the fenfluramine tablets being labelled as placebo. All three groups of subjects were found to reduce their food intake at lunch in their own homes by about 30 per cent, regardless of the tablet contents, so long as the latter were marked as an appetite suppressant. However, when fed in the laboratory from a food-dispensing machine neither tablet label nor drug affected the food intake of the obese groups, although the normal-weight subjects reduced their calorie intake by 23 per cent on fenfluramine.

Effects of fenfluramine (40 mg) on salivation and hunger ratings were studied by Wooley, Wooley, Williams, and Nurre 1977), they found no effect of the drug on either of these measurements after one hour. However, we have recently shown that a 60 mg dose, but not a 30 mg dose, of fenfluramine depressed hunger ratings three and seven hours after ingestion of the drug compared with placebo. Fenfluramine also reduced both resting salivation and the salivary response to food, the 60 mg dose being significantly more potent than 30 mg of this drug or 10 mg amphetamine. Calorie intake was lower with fenfluramine than with placebo, 8 per cent and 26 per cent with 30 mg and 60 mg respectively, compared with a 21 per cent reduction of food intake with amphetamine in the same study. Furthermore, changes in 'hunger', 'desire to eat' and a subjective estimate of likely food consumption, were highly correlated with changes in actual calorie intake with fenfluramine (Rogers and Blundell 1979). Hunger ratings produced by fenfluramine have been tracked for up to eight hours (Silverstone and Kyriakides 1982). 20, 40 and 80 mg suppressed hunger over this time period in a dose-related manner.

Silverstone (1972) has also reported that phentermine reduces food intake (by 25 per cent and 30 per cent over two meals with 15 mg and 30 mg phentermine resinate respectively) in overweight female subjects. Hunger ratings showed a corresponding reduction when summed over the two meals. In clinical trials other authors earlier reported mild appctite-reducing effects amongst the subjective effects of phentermine and chlorphentermine. Mazindol, in one study, did not decrease stimulus-induced salivation, but overweight subjects receiving this drug noted less hunger prior to eating than with placebo (Johnson and Hughes 1977). With tiflorex, both hunger scores and food intake, measured using an automated food dispenser for six normal-weight subjects, were reduced compared with placebo. The effect was maximal after five to seven hours, and in a comparative study 20 mg tiflorex was found to have a greater anorectic effect than diethylpropion (75 mg), fenfluramine (60 mg), mazindol (2 mg) and phentermine (60 mg) (Fincham and Silverstone 1977). In an interesting study with phenmetrazine normal subjects reduced their food intake with drug compared with a placebo only when they were told that appetite suppression might occur. No effects on hunger, scored on a 'symptom checklist' were seen (Penick and Hinkle 1964). Phenylpropanolamine, the active ingredient in most non-prescription weight-loss products sold in the US, suppressed lunchtime food intake in weight-conscious volunteers in two studies, but not in a third using subjects motivated only by pay (Hoebel, Hamilton and Kornblith 1977).

Several cen*t*rally-acting anorectic drugs have been shown to reduce the food consumption of both obese and normal-weight persons in the laboratory under double-blind protocol. Hunger ratings obtained using visual analogue scales may be similarly reduced. Weak, but positive, correlations between change in hunger ratings and reduction in calorie intake tentatively suggest that the primary action of at least some of these drugs is to reduce the motivation to eat (hunger) with a consequent secondary reduction in food intake. On the other hand, certain drugs may exert a non-specific action, and divert attention toward side-effects, and away from eating and its psychological correlates.

In addition, the subjective effects of the drugs are variable and hunger suppression does not occur in all subjects nor under all experimental circumstances. These findings suggest that effects on perceived hunger may be markedly influenced by expectations and by the 'demand characteristics' of the investigations.

C. Mechanisms of action

Whilst the efficacy of centrally-acting drugs to adjust hunger sensations and food consumption have been amply demonstrated in man, the mechanisms underlying such changes have usually been imputed from animal studies. It is clearly more difficult to investigate neurochemical mechanisms and biological processes in man. However, the involvement of catecholamines in the anorectic action of amphetamine (see Section 00) has received attention. The strategy was designed to reveal a role for either dopamine or noradrenaline. Interestingly, pimozide – a drug which blocks dopamine receptors – antagonized the effect of *d*-amphetamine on arousal but exerted no effect on feelings of hunger. On the other hand the noradrenaline receptor-blocking drug thymoxamine did appear to antagonize the amphetamine hunger effect in a dose-dependent manner (Silverstone and Kyriakides 1982). This effect is consistent with the data from animal studies using low doses of amphetamine (Blundell and Burridge 1979).

At a different level of analysis the mechanism of action of drugs can be defined according to the process through which food intake is reduced. That is by a primary action on hunger, appetite or satiety (Blundell 1979). Although drugs are unlikely to exert pure effects on these processes it has been suggested on the basis of animal studies that amphetamine acts primarily by a suppression of hunger whilst fenfluramine promotes the onset of satiation. In turn, these actions are linked to the roles of catecholamines and serotonin in the control of food intake (Blundell, Latham, and Leshem 1976; Blundell 1979). In man it is only possible to ascribe such functions to the action of drugs if the qualitative structure of feeding behaviour can be analysed. Accordingly, it is interesting that in one study which employed the videotaping of a test meal, amphetamine and fenfluramine displayed quite distinctive actions on the eating profile. The slow rate of eating engendered by fenfluramine (Rogers and Blundell 1979) together with subsequent intensification of feelings of stomach fullness (Blundell and Rogers 1980) are quite consistent with a major effect of fenfluramine on the process of satiation and the state of satiety. On the other hand, although the effects of amphetamine were ambiguous, the data suggested that this drug acted, in part, by inhibiting the process of hunger (that is, by blocking the tendency to initiate eating).

D. Importance of hunger suppression

All centrally-acting drugs, whether they act through the processess of hunger or satiation, will give rise to changes in reported subjective feelings of hunger (note that the conscious sensation of hunger should not be confused with the objectively defined process called 'hunger motivation'). On many occasions however the intensity of reported hunger is only weakly correlated with the amount of food subsequently consumed. Indeed, in certain cases such as following the administration of naloxone food intake may be significantly lowered with no effect on subjective hunger (Trenchard and Silverstone 1983). This finding indicates that hunger and food intake can be dissociated. What therefore is the relationship between hunger and food intake following anorectic drug administration? The traditional interpretation is that anorectic drugs reduce subjective perception of hunger which in turn leads to a reduction in food consumption. However, the correlation between hunger and food intake is based on a single measure of each variable. Normally, hunger is evaluated before the start of a test meal and subsequently correlated with the amount of food consumed during the meal. However, in a recent study hunger was measured not only before a test meal but at various stages during the course of consumption. This procedure is called 'temporal tracking'. This technique demonstrated that although the anorectic drugs amphetamine and fenfluramine reduced hunger before the start of the meal, as soon as eating began hunger was reinstated to control levels (Fig. 4.5) and remained there for the duration of the meal (Blundell, Brownlie, and Rogers 1979). This finding suggests that following drug administration

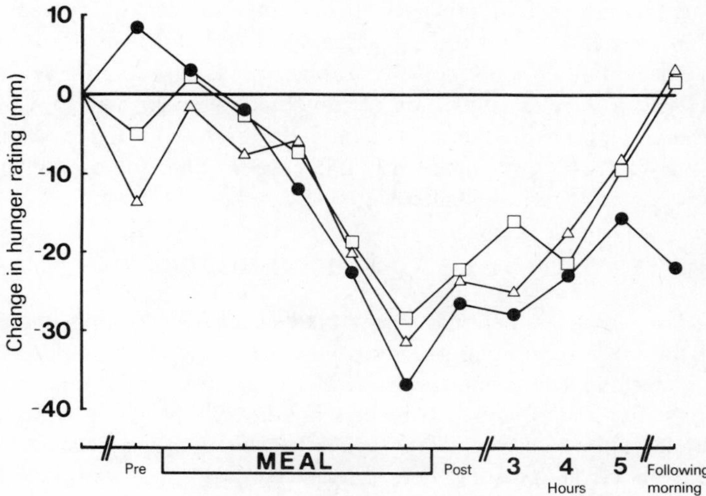

FIG. 4.5 Profile of hunger ratings following drug administration (•, placebo; □, amphetamine; △, fenfluramine). Each point represents the change score from pre-drug baseline. Note that the abscissa is not an equal interval scale but indicates the times at which ratings were made.

hunger may play little part in controlling food consumption once eating has begun. In turn this may explain the frequently-observed low correlation between hunger and amount eaten and suggests that the inhibition of hunger may not be the major reason why subjects eat less following drug administration. The relationship between subjective feelings of hunger and food intake appears to involve a complex interaction of physiological and psychological elements.

E. Macronutrient selection

Animal studies have suggested that certain centrally-acting drugs with actions on specific neurotransmitters should modulate selection and consumption of protein and carbohydrate. Of particular interest are those drugs which adjust the synaptic activity of serotonin though it should be noted that not all studies are in agreement (see Section 8.E). However, using amphetamine and fenfluramine as a drug pair, in a manner similar to that used in animal studies, it was reported that amphetamine produced a marked suppression of protein consumption whereas fenfluramine exerted a weak suppression of carbohydrate intake (Rogers and Blundell 1979). Similar selective effects were observed on a food preference questionnaire, and it was noticeable that the subjective preferences were a more sensitive indicator of the presence of an active drug than actual behaviour. In a longer term study a specific test was made of the capacity of serotoninergic agents to suppress carbohydrate intake. A group of obese subjects, prone to snack on high carbohydrate foods, were treated separately with tryptophan and fenfluramine (Wurtman and Wurtman 1981; Wurtman, Wurtman, Growdon, Henry, Lipscomb and Zeisel 1982). The treatment period for each compound was five days, and during this time fenfluramine significantly reduced the number of carbohydrate foods taken in snacks. Tryptophan gave rise to a weaker effect. This finding may prove useful in the treatment of carbohydrate cravers but does not constitute strong proof of a role for serotonin in nutrient control since the inhibition of carbohydrate intake may have been part of a general suppressive effect on eating. However, evidence is available to suggest that centrally-acting agents may influence the direction of food choice as well as the amount consumed (Blundell 1983).

10. CURRENT TRENDS AND FUTURE DIRECTIONS

Over the last few years there has been a marked tendency in animal research to broaden the range of feeding situations in which anorectic drugs are tested. A number of techniques for inducing hyperphagia are now available in addition to food deprivation, and the use of these techniques will provide a richer pool of information from which to evaluate the mechanism of drug action and to anticipate the likely effect in man. In human research, measuring devices have become more sophisticated. In addition to the use of hunger-rating scales and measures of total calorie intake, researchers have used various automated devices, in addition to video-recording, for analysing moment-to-moment changes in eating after drug administration. These techniques have provided information about the action of drugs on processes of hunger, appetite or satiation. As well as total quantity of food ingested, the direction of motivation has been monitored through food preferences and selection, with particular emphasis on protein and carbohydrate intake. The last 10 years have seen significant developments in the analysis of anorectic drug action in animals and man.

At the present time the major target for investigations of the mechanisms of action of anorectic drugs remains the biogenic amines. Current accounts of drug action are explained with reference to dopaminergic, adrenergic, or serotoninergic systems. However, there are indications that this will broaden to include the opioids, peptides, and GABA. Indeed there already exists a theoretical scheme which draws together many of these newer findings (Morley and Levine 1983).

Regarding the use of anorectic drugs in clinical practice, these compounds remain the most widely used in the treatment of obesity. However, owing to the risk of drug abuse that may be associated with amphetamine-like compounds, emphasis in research has shifted from central to peripheral sites of action, and from the control of food intake to the modification of energy output. At the moment it is too early to evaluate the success of these alternative lines of research. Thermogenic compounds are now at the stage of clinical trials and their impact on the treatment of obesity should be known within a few years. Whatever the outcome of this enterprise it is likely that drugs influencing appetite and satiation, or modifying food preferences, will continue to have a role in the treatment of eating disorders. It seems certain that drugs will be used more selectively, with compounds displaying particular pharmacological and behavioural profiles being tailored to the requirements of individuals with specific types of eating problems. Perhaps drugs for the treatment of obesity, whether directed primarily toward the adjustment of metabolism or behaviour, can be used to give patients an opportunity to learn something about themselves. Whatever programme is analysed for the treatment of obesity, people will always have to eat, and eating is an activity which links the biological domain of energy transactions with the psychological domain of knowledge. Thus, in experimental settings, drugs can be used as tools to further explore the processes underlying hunger and feeding, while in the clinic drugs could be used as tools to assist people to better understand the forces - biological and psychological - which control their behaviour.

References

Ahlskog, J. E. (1974). Food intake and amphetamine anorexia after selective forebrain norepinephrine loss. *Brain Res.* **82**, 211–40.

Anderson, G. H. (1977). Regulation of protein intake by plasma amino acids. *Adv. Nutrit. Res.* **1**, 145–66.

—— (1979). Control of protein and energy intake: role of plasma amino acids and brain neurotransmitters. *Can. J. Physiol. Pharmac.* **57**, 1043–57.

Antelman, S. M., Caggiula, A. R., Black, C. A., and Edwards, D. J. (1978). Stress reverses the anorexia induced by amphetamine and methylphenidate but not fenfluramine. *Brain Res.* **143**, 580–5.

Apfelbaum, M. and Mandenoff, A. (1981). Naltrexone suppresses hyperphagia induced in the rat by a highly palatable diet. *Pharmacol. Biochem. Behav.* **15**, 89–91.

Ashley, D. V. M. and Anderson, G. H. (1975). Food intake regulation in the weanling rat: effects of the most limiting essential amino acids of gluten, casein and zein on the self-selection of protein and energy. *J. Nutrit.* **105**, 1405–11.

—— Coscina, D. V., and Anderson, G. H. (1979). Selective decrease in protein intake following brain serotonin depletion. *Life Sci.* **24**, 973–84.

Baez, L. A. (1974). Role of catecholamines in the anorexic effects of amphetamine in rats. *Psychopharmacology* **35**, 91–6.

Bahnsen, P., Jacobson, E., and Thesleff, H. (1938). The subjective effect of beta-phenylisopropylaminesulfate on normal adults. *Acta med. scand.* **97**, 98–131.

Balopole, D. C., Hansult, C. D., and Dorph, D. (1979). Effect of cocaine on food intake in rats. *Psychopharmacology* **64**, 121–2.

Beaudoin, R, and Mayer, J. (1953). Food intake of obese and non-obese women *J. Am. Diet. Assoc.* **29**, 29–33.

Bedford, J. A., Lovell, D. K., Turner, C. E., Elsohly, M. A., and Wilson, M. C. (1980). The anorexic and actometric effects of cocaine and the coca extracts. *Pharmacol. Biochem. Behav.* **13**, 403–8.

Bellisle, F. and Le Magnen, J. (1980). The analysis of human feeding patterns: the edogram. *Appetite* **1**, 141–50.

Bernstein, C. M. and Grossman, M. I. (1956). An experimental test of the glucostatic theory of regulation of food intake. *J. Clin. Invest.* **35**, 626–33.

Biel, J. H. (1970). Structure–activity relationships of amphetamine and derivatives. In *Amphetamine and related compounds* (eds. E. Costa and S. Garrattini) pp. 3–19. Raven Press, New York.

Blundell, J. E. (1975). Anorexic drugs, food intake and the study of obesity. *Nutrition (Lond.)* **29**, 5–18.

—— (1977). Is there a role for serotonin (5-hydroxytryptamine) in feeding? *Int. J. Obes.* **1**, 15–42.

—— (1979*a*). Hunger, appetite and satiety – constructs in search of identities. In *Nutrition and lifestyles* (ed. M. Turner) pp. 21–42. Applied Science Publishers, London.

—— (1979*b*). Serotonin and feeding. In *Serotonin in health and disease* (ed. W. B. Essman) Vol. 5, Clinical applications, pp. 403–50. Spectrum, New York.

—— (1980). Pharmacological adjustment of the mechanism underlying feeding and obesity. In *Obesity* (ed. A. J. Stunkard) pp. 182–207. Saunders, Philadelphia.

—— (1981). Biogrammar of feeding: pharmacological manipulations and their interpretations. In *Progress in theory in psychopharmacology* (ed. S. J. Cooper) pp. 233–76. Academic Press, London.

—— (1982). Neuroregulators and feeding: implications for the pharmacological manipulation of hunger and appetite. *Rev. Pure Appl. Pharmacol. Sci.* **3(4)**, 381–462.

—— (1983). Processes and problems underlying the control of food selection and nutrient intake. In *Nutrition and the brain* (eds. R. J. Wurtman and J. J. Wurtman) Vol. 6, pp. 163–221. Raven Press, New York.

—— and Burridge, S. L. (1979). Control of feeding and the psychopharmacology of anorexic drugs. In *The treatment of obesity* (ed. J. Munro) pp. 53–84. M.T.P. Press, Lancaster.

Blundell, J. E. and Latham, C. J. (1977). Pharmacolgical modification of eating behaviour. *Proc. 6th int. cong. physiology of food and fluid intake*, Paris.

—— —— (1978). Pharmacological manipulation of feeding behaviour: possible influences of serotonin and dopamine on food intake. In *Central mechanisms of anorectic drugs* (eds. S. Garrattini and R. Samanin) pp. 83–109. Raven Press, New York.

—— —— (1979*a*). Pharmacology of food and water intake. In *Chemical influences on behaviour* (eds. S. Cooper and K. Brown) pp. 201–54. Academic Press, London.

—— —— (1979*b*). Serotonergic influences on food intake: effect of 5-hydroxytryptophan on parameters of feeding behaviour in deprived and free-feeding rats. *Pharmac. Biochem. Behav.* **11**, 431–7. Raven Press, New York.

—— —— (1980). Characterization of the adjustments to the structure of feeding behaviour following pharmacological treatment: effects of amphetamine and fenfluramine and the antagonism produced by pimozide and methergoline. *Pharmac. Biochem. Behav.* **12**, 717–22.

—— —— (1982). Behavioural pharmacology of feeding. In *Drugs and appetite* (ed. T. Silverstone) pp. 41–80. Academic Press, London.

—— and Leshem, M. B. (1975). Analysis of the mode of action of anorexic drugs. In *Recent advances in obesity research* **Vol. I** (ed. A. Howard) pp. 368–71. Newman, London.

—— and McArthur, R. A. (1979). Investigation of food consumption using a dietary self-selection procedure: effects of pharmacological manipulation and feeding schedules. *Br. J. Pharmac.* **67**, 436–8.

—— and Rogers, P. J. (1978). Pharmacological approaches to the understanding of obesity. *Psychiatric Clinics of N. America* **1**, 629–50.

—— —— (1980*a*). Effects of anorexic drugs on food intake, food selection and preferences, hunger motivation and subjective experiences: Pharmacological manipulation as a tool to investigate human feeding processes. *Appetite* **1**, 151–65.

—— —— (1980*b*). Fame e appetito – una prospettiva biopsicologica. In *Obesita* (eds. M. Cairella and A. Jacobelli) pp. 99–119. Societa Editrice Universo, Roma.

—— Brownlie, V., and Rogers, P. J. (1979). Hunger and eating: a paradoxical effect of anorexic drugs in man. *Int. J. Obes.* **3**, 399–400.

—— Cambell, D. B., Leshem, M. B., and Tozer, R. (1975). Comparison of the time course of the anorexic effects of amphetamine and fenfluramine with drug levels in blood. *J. Pharm. Pharmacol.* **27**, 187–92.

Blundell, J. E., Latham, C. J., and Leshem, M. B. (1973). Biphasic action of a 5-hydroxytryptamine inhibitor on fenfluramine induced anorexia. *J. Pharm. Pharmac.* **25**, 492–4.

Borbely, A. and Waser, P. G. (1966). Das fessverhalten der Ratte. Haufigkeit und Gewicht der Mahlzeiten unter dem Einfluss von Amphetamin. *Psychopharmacologia* **9**, 373–81.

Borsini, F., Bendotti, C., Aleotti, A., Samanin, R., and Garattini, S. (1982). *D*-Fenfluramine and *D*-norfenfluramine reduce food intake by acting on different serotonin mechanisms in the rat brain. *Pharmac. Res. Comm.* **14**, 671–8.

—— —— Carli, M., Poggesi, E., and Samanin, R. (1979). The roles of brain noradrenaline and dopamine in the anorectic activity of diethylpropion in rats: a comparison with *d*-amphetamine. *Res. Comm. Chem. Path. Pharmacol.* **26**, 3–11.

—— —— Thurlby, P., and Samanin, R. (1982). Evidence that systemically administered salbutamol reduces food intake in rats by acting on central beta-adrenergic sites. *Life Sci.* **30**, 905–11.

Bowden, C., White, K. and Tutwiler, G. (1983). Inhibition of energy intake of cafeteria fed rats by mechanistically different anorectic agents. In *Proc. 4th. Cong. Obesity* Oct. 5–8, New York. p.35A.

Bray, G. A. and York, D. A. (1972). Studies on food intake in genetically obese rats. *Am. J. Physiol.* **233**, 176–9.

Burton, M. J., Cooper, S. J., and Popplewell, D. A. (1981). The effect of fenfluramine on the microstructure of feeding and drinking in the rat. *Br. J. Pharmac.* **72**, 621–33.

Burridge, S. L. and Blundell, J. E. (1979). Amphetamine anorexia: antagonism by typical but not atypical neuroleptics. *Neuropharmacology* **18**, 453–7.

Carey, R. J. (1976). Effects of selective forebrain depletions of norepinephrine and serotonin on the activity and food intake effects of amphetamine and fenfluramine. *Pharmac. Biochem. Behav.* **5**, 519–23.

Carruba, M. O., Mantegazza, P., Muller, E. E. and Ricciardi, S. (1981). Effects of anorectic drugs on the hyperphagic response induced by insulin or 2-deoxy-d-glucose glucoprivation. *Br. J. Pharmac.* **72**, 161.

—— Ricciardi, S., and Mantegazza, P. (1980*a*). Reduction of food intake by piribidel in the rat: relation to dopamine receptor stimulation. *Life Sci.* **27**, 1131–40.

—— Ricciardi, S., Muller, E. E., and Mantegazza, P. (1980*b*). Anorectic effect of lisuride and other ergot derivatives in the rat. *Eur. J. Pharmac.* **64**, 133–41.

—— —— Muller, E. E., and Mantegazza, P. (1980*c*). New findings in the neuropharmacological control of food intake. *Pharmacol. Res. Comm.* **12** 599–603.

Clineschmidt, B. V., McGuffin, J. C., and Werner, A. B. (1974). Role of monoamines in the anorexigenic actions of fenfluramine amphetamine and *p*-chloroamphetamine. *Eur. J. Pharmac.* **27**, 313–23.

Cole, S. O. (1978). Brain mechanisms of amphetamine-induced anorexia, locomotion and stereotypy: a review. *Neurosci. Biobehav. Rev.* **2**, 89–100.

—— and Gay, P. E. (1974). Brain mechanisms underlying the effects of amphetamine on feeding and non-feeding behaviours: dissociation and overlap. *Physiol. Psychol.* **2**, 80–8.

Collier, G., Leshner, A. I., and Squibb, R. L. (1969). Dietary self-selection in active and non-active rats. *Physiol. Behav.* **4**, 79–82.

Cooper, B. R., Howard, J. L., White, H. L., Soroko, F., Ingold, K., and Maxwell, R. A. (1980). Anorexic effect of ethanolamine-o-sulphate and muscimol in the rat: evidence that GABA inhibits ingestive behaviour. *Life Sci.* **26**, 1997–2002.

Corey, D. T., Walton, A., and Wiener, N. I. (1978). Development of carbohydrate preference during water rationing: a specific hunger? *Physiol. Behav.* **20**, 547–52.

Costa, E. and Garattini, S. (1970). Amphetamines and related compounds. Raven Press, New York.

Cox, R. H. and Maickel, R. D. (1972). *J. Pharmacol. Exp. Ther.* **181**, pp. 1–00.

Creese, I. and Iversen, S. D. (1975). The pharmacological and anatomical substrates of the amphetamine response in the rat. *Brain Res.* **83**, 419–36.

Davidoff, E. and Reifenstein, E. C. (1937). The stimulating action of benzedrine sulphate: a comparison study of the responses of normal persons and depressed patients. *JAMA* **108**, 1770–6.

Davies, R. F. (1976). Some neurochemical and physiological factors controlling free feeding patterns in the rat. *Ph.D. thesis*, McGill University, Montreal.

—— Rossi, J., Panksepp, J., Bean, N. J., and Zolovick, A. J. (1983). Fenfluramine anorexia: a peripheral locus of action. *Physiol. Behav.* **30**, 723–30.

Davis, J. D., Gallagher, R. L., and Ladove, R. (1967). Food intake controlled by a blood factor. *Science* **156**, 1247–8.

Dobrzanski, S. and Doggett, N. S. (1976). The effects of (+)- amphetamine and fenfluramine on feeding in starved and satiated mice. *Psychopharmacology* **48**, 283–6.

Doggett, N. S. and Jawaharlal, K. (1977*a*). Some observations on the anorectic activity of prostglandin $F_{2\alpha}$. *Br. J. Pharmac.* **60**, 409–15.

—— —— (1977*b*). Anorectic actvity of prostaglandin precursors. *Brit. J. Pharmac.* **60**, 417–28.

Eichler, A. J. and Antelman, S. M. (1977). Apomorphine: feeding or anorexia depending on internal state. *Commun. Psychopharm.* **1**, 533–40.

Fahrbach, S. E., Tretter, J. R., Aravich, P. F., McCabe, J., and Leibowitz, S. F. (1980). Increased carbohydrate preference in the rat after injection of 2-d eoxy-d-glucose and clonidine. In *Proc. Soc. for Neurosci. 10th annual meeting*, Abstract, Vol. 6, p. 784.

Fernstrom, J. D. and Wurtman, R. J. (1972). Brain serotonin content: physiological regulation by plasma neutral amino acids. *Science* **178**, 414–6.

—— —— (1973). Control of brain 5-HT content by dietary carbohydrates. In *Serotonin and behaviour* (eds. J. Barchas and E. Usdin) pp. 121–8. Academic Press, New York.

Fincham, J. and Silverstone, J. T. (1977). The anorectic effect of triflorex, a new appetite suppressant compound. In *Proc. 2nd Int. Cong. Obesity* (Washinton, D. C.), Abstract. p.6.

Freed, W. J., Perlow, M. J. and Wyatt, R. J. (1979). Calcitonin: inhibitory effect on eating in rats. *Science* **206**, 850–2.

Fuxe, K., Farnebo, L. O., Hamberger, B., and Ogren, S. O. (1975). On the *'in vivo'* and *'in vitro'* actions of fenfluramine and its derivatives on central monoamine neurones, especially 5-hydroxytryptamine neurones, and their relation to the anorectic action of fenfluramine. *Postgrad. Med. J.* **51**, 1, 35–45.

Garfinkel, P. E. (1974). Perception of hunger and satiety in anorexia nervosa. *Psychol. Med.* **4**, 309–15.

Garattini, S. (1978). Importance of serotonin for explaining the action of some anorectic agents. In *Recent advances in obesity research: II*, (ed. G. A. Bray), 433–41. Newman, London.

—— and Samanin, R. (1976). Anorectic Drugs and Neuro-transmitters. In *Food intake and appetite* (ed. T. Silverstone) pp. 82–108. Dahlem Konferenzen, Berlin.

—— —— (1978). Amphetamine and fenfluramine, two drugs for studies on food intake. *Int. J. Obes.* **2**, 349–51.

—— Caccia, T., Mennini, T., Samanin, R., Consolo, S., and Ladinsky, H. (1979). Biochemical pharmacology of the anorectic drug fenfluramine: a review. *Current medical research and opinion* **6**, Suppl. 1, 15–23.

Gibbs, J., Young, R. C., and Smith, G. P. (1973). Cholecystokinin elicits satiety in rats with open gastric fistulas. *Nature* **245**, 323–5.

Glick, S. D., Walters, D. H. and Milloy, S. (1973). Depletion of hypothalamic norepinephrine by food deprivation and interaction with *d*-amphetamine. *Res. Commun. Chem. Path. Pharmac.* **6**, 773–8.

Glick, Z. (1980). Food intake of rats administered with glycerol. *Physiol. Behav.* **25**, 621–6.

Grinker, J. A., Drewnowski, A., Enns, M. and Kissileff, H. (1980). Effects of *d*-amphetamine and fenfluramine on feeding patterns and activity of obese and lean Zucker rats. *Pharmac. Biochem. Behav.* **12**, 265–75.

Harris, S. C., Ivy, A. C., and Searle, C. M. (1974). The mechanisms of amphetamine-induced loss of weight. *JAMA* **134**, 1468–75.

Hashim, S. A. and Van Itallie, T. B. (1964). An automatically monitored food dispensing apparatus for the study of food intake in man. *Fed. Proc. Fedn. Am. Socs. Exp. Biol.* **23**, 82.

Hill, A. J. and Blundell, J. E. (1982/83). Nutrients and behaviour: research strategies for the investigation of taste characteristics, food preferences, hunger sensations and eating patterns in man. *J. Psychiat. Res.* **17**, 203–12.

—— Rogers, P., and Blundell, J. (1983). Effect of an anorexic drug (d-fenfluramine) on body weight and food intake during the dynamic and plateau phases of dietary-induced obesity in rats: more potent drug action in obese animals. In *Proc. 4th Int. Cong. Obesity*, (New York) Oct. 5–8. p. 114.

Hoebel, B. G. (1974). Brain reward and aversion systems in the control of feeding and sexual behaviour. *Neb. Symp. Motiv.* **22**, 49–122.

—— (1978). Three anorectic drugs: similar structures but different effects on brain and behaviour. *Int. J. Obes.* **2**, 157–66.

—— (1984). Neurotransmitters in the control of feeding and its rewards: monoamines, opiates and brain-gut peptides. In *Eating and its disorders* (eds. Stunkard, A. J. and Stellar, E.). Raven Press, New York.

—— and Leibowitz, S. F. (1981). Brain monoamines in the modulation of self-stimulation, feeding, and body weight. In *Brain, behaviour and bodily disease*, (eds. H. A. Weiner, M. A. Hofner and A. J. Stunkard) pp. 103–42. Raven Press, New York.

—— Hamilton, C. C., and Kornblith, C. L. (1977). Human, monkey and rat studies of the O.T.C. anorectic drug, phenylpropanolamine. *Proc. 2nd Int. Cong. Obesity*, (Washington D. C.) Abstract p.33.

—— Hernandez, L., McLean, S., Stanley, B. G., Aulissi, E. F., Glimcher, P., and Margolin, D. (1982). Catecholamines, enkephalin and neurotensin in feeding and reward. In *The neural basis of feeding and reward* (eds. B. G. Hoebel and D. Novin) pp. 465–78. Haer Institute, Brunswick.

Hollister, L. E. (1971). Hunger and appetite after single doses of marihuana, alcohol and dextroamphetamine. *Clin. Pharmac. Ther.* **12**, 44–9.

Holtzman, S. G. (1974). Behavioural effects of separate and combined administration of naloxone and *d*-amphetamine. *J. Pharmac. Exp. Ther.* **189**, 51–60.

Jacobson, E. (1939). Studies on the subjective effects of the cephalotropic amines in man II. *Acta med. scand.* **100**, 188–201.

—— and Wortsein, A. (1939). Studies on the subjective effects of the cephalotropic amines in man I. *Acta med. scand.* **100**, 159–87.

Jespersen, S. and Scheel-Kruger, J. (1973). Evidence for a difference in mechanism of action between fenfluramine- and amphetamine-induced anorexia. *J. Pharm. Pharmac.* **25**, 49–54.

Johnson, W. G. and Hughes, J. R. (1977). The evaluation of mazindol in the treatment of obesity. *Proc. 2nd Int. Cong. Obesity* (Washington D. C.) (Abstract) p. 6.

Jordan, H. A., Weiland, W. F., Zebley, S. P., Stellar, E., and Stunkard, A. J. (1966). Direct measurement of food intake in man: a method for the objective study of eating behaviour. *Psychosom. Med.* **28**, 836–42.

Kanarek, R. B. and Beck, J. M. (1980). Role of gonadal hormones in diet selection and food utilization in female rats. *Physiol. Behav.* **24**, 381–6.

—— Ho, L., and Meade, R. G. (1981). Amphetamine selectively decreases fat consumption in rats. *Pharmac. Biochem. Behav.* **14**, 539–42.

—— Marks-Kaufman, R. and Lipeles, B. J. (1980). Increased carbohydrate intake as a function of insulin administration in rats. *Physiol. Behav.* **25**, 779–82.

Kirby, M. J., Pleece, S. A., and Redfern, P. H. (1978). The effect of fenfluramine on obesity in rats – a new method for the screening of potential anti-obesity agents. *Br. J. Pharmac.* **64**, 442.

Kissileff, H. R., Klingsberg, G., and Van Itallie, T. B. (1980). Universal eating monitor for continuous recording of solid or liquid consumption in man. *Am. J. Physiol.* **238**, R14–R22.

—— Pi-Sunyer, F. X., Thornton, J., and Smith, G. P. (1981). *C*-terminal octapeptide of cholecystokinin decreases food intake in man. *Am. J. Clin. Nutr.* **34**, 154–60.

Knoll, J. (1979). Satietin: A highly potent anorexigenic substance in human serum. *Physiol. Behav.* **23**, 497–502.

Kroger, W. S. (1962). A comparison of anorexic drugs in the treatment of the resistant obese patient. *Psychosomatics* **3**, 454–57.

Latham, C. J. and Blundell, J. E. (1979). Evidence for the effect of tryptophan on the pattern of food consumption in free feeding and food deprived rats. *Life Sci.* **24**, 1971–8.

Leibowitz, S. F. (1978). Paraventricular nucleus: a primary site mediating adrenergic stimulation of feeding and drinking. *Pharmac. Biochem. Behav.* **8**, 163–75.

—— and Brown, L. L. (1980*a*). Histochemical and pharmacological analysis of noradrenergic projections to the paraventricular hypothalamus in relation to feeding stimulation. *Brain Res.* **201**, 289–314.

—— —— (1980*b*). Histochemical and pharmacological analysis of catecholaminergic projections to the perifornical hypothalamus in relation to feeding inhibition. *Brain Res.* **201**, 315–45.

—— and Hor, L. (1980). Behavioural effect of beta-endorphin (β-EP) and norepinephrine in the hypothalamic paraventricular nucleus (PNV) *Neurosci. Abstr.* **6**, 318.

—— and Papadakos, P. J. (1978). Serotonin–norepinephrine interactions in the paraventricular nucleus: antagonistic effects of feeding behaviour in the rat. *Neurosci. Abstr.* **4**, 542.

—— and Rossakis, C. (1979). Analysis of feeding suppression produced by periforhypothalamic injection of catecholamines, amphetamines and mazindol. *Eur. J. Pharmac.* **53**, 69–81.

—— Arcomano, A. and Hammer, N. J. (1978). Potentiation of eating associated with tri-cyclic antidepressant drug activation of α–adrenergic neurones in the paraventricular nucleus. *Prog. Neuropychopharmacol.* **2**, 349–58.

Leshner, A. I., Collier, G. H. and Squibb, R. L. (1971). Dietary self-selection at cold temperatures. *Physiol. Behav.* **6**, 1–3.

—— Siegel, H. I., and Collier, G. (1972). Dietary self-selection by pregnant and lactating rats. *Physiol. Behav.* **8**, 151–4.

Levin, B. Thermogenic agents as a treatment of obesity. (1983) *Proc. 4th Int. Cong. Obesity,* New York, 5–9 Oct. p. 00.

Lorenz, D., Nardi, P. and Smith, G. P. (1978). Atropine methyl nitrate inhibits sham feeding in the rat. *Pharmac. Biochem. Behav.* **8**, 405–7.

Lotter, E. C., Krinsky, R., McKay, J. M., Treneer, C. M., Porte, D. and Woods, S. C. (1981). Somatostatin decreases food intake of rats and baboons. *J. Comp. Physiol. Psychol.* **95**, 278–87.

Lyon, M. and Robbins, T. (1975). The action of central nervous system stimulant drugs: a general theory concerning amphetamine effects. In *Current Developments in Psychopharmacology* (eds. W. V. Essman and L. Valzelli) Vol. II. Spectrum, New York.

McArthur, R. A. and Blundell, J. E. (1983). Protein and carbohydrate self-selection: modification of the effects of fenfluramine and amphetamine by age and feeding regimen. *Appetite* **4**, 113–24.

Malcolm, A. D., Mace, P. M., Ontar, K. P. *et al.* (1972). Experimental evaluation of anorexic agents in man: a pilot study. *Proc. Nutr. Soc.* **31**, 12A.

Mandenoff, A., Fumeron, F., Apfelbaum, M., and Margules, D. L. (1982). endogenous opiates and energy balance. *Science* **215**, 1536–8.

Marks-Kaufman, R., and Kanarek, R. B. (1980). Morphine selectively influences macronutrient intake in the rat. *Pharmac. Biochem. Behav.* **12**, 427–30.

Martin, C. F. and Gibbs, J. (1980). Bombesin elicits satiety in sham feeding rats. *Peptides* **1**, 131–4.

Mauron, C., Wurtman, J. J., and Wurtman, R. J. (1980). Clonidine increases food and protein consumption in rats. *Life Sci.* **27**, 781–91.

Monello, L. F. and Mayer, J. (1967). Hunger and satiety sensations in men, women, boys and girls. *Am. J. Clin. Nutr.* **20**, 253–61.

Moon, R. D. (1979). Monitoring human eating patterns during the ingestion of non-liquid foods. *Int. J. Obes.* **3**, 281–8.

Morley, J. E. (1980). The neuroendocrine control of appetite: the role of the endogenous opiates, cholecystokinin, T.R.H., gamma-animo-butyric-acid and the diazepam receptor. *Life Sci.* **27**, 355–78.

—— and Levine, A. S. (1983). The central control of appetite. *Lancet* 398–401.

Nathanson, M. H. (1937). The central action of beta-aminopropylbenzene (Benzedrine). *JAMA* **108**, 528–31.

Neckers, L. M., Biggio, G., Moja, E., and Meek, J. L. (1977). Modulation of brain tryptophan hydroxylase activity by brain tryptophan content. *J. Pharmac. Exp. Ther.* **201**, 110–6.

Norton, S. (1973). Amphetamine as a model for hyperactivity in the rat. *Physiol. Behav.* **11**, 181–6.

Overmann, S. R. (1976). Dietary self-selection by animals. *Psycholog. Bull.* **83**, 218–35.

—— and Yang, M. G. (1973). Adaptation to water restriction through dietary selection in weanling rats. *Physiol. Behav.* **11**, 781–6.

Paul, S. M., Hulihan-Giblin, B., and Skolnick, P. (1982). (+)– Amphetamine binding to rat hypothalamus: relation to anorexic potency of phenylethylamines. *Science* **218**, 487–90.

Penick, S. B. and Hinkle, L. E. (1964). The effect of expectation on response to phenmetrazine. *Psychosomat. Med.* **26**, 369–73.

Peters, G., Besseghir, K. P., Kaserman, H-P., and Peters–Haefeli, L. (1979). Effects of drugs on ingestive behaviour. *Pharmac. Ther.* **5**, 485–503.

Pudel, V. E. (1975). Psychological observations on experimental feeding in the obese. In *Recent advances in obesity research* I. (ed. A. N. Howard) Vol. I, pp. 217–20. Libby, London.

—— and Oetting, M. (1977). Eating in the laboratory: behavioural aspects of the positive energy balance. *Int. J. Obes.* **1**, 369–86.

Quattrone, A., Bendotti, C., Recchia, M., and Samanin, R. (1977). Various effects of d-amphetamine in rats with selective lesions of brain noradrenaline-containing neurones or treated with perfluridol. *Commun. Psychopharm.* **1**, 525–31.

Richter, C. P. (1943). Total self-regulatory functions in animals and human beings. *Harvey Lecture Series* **38**, 63–103.

Rogers, P. J. and Blundell, J. E. (1979). Effect of anorexic drugs on food intake and the micro-structure of eating in human subjects. *Psychopharmacology* **66**, 159–65.

Rozin, P. (1976). The selection of foods by rats, humans and other animals. In *Advances in the study of behaviour,* (eds. J. Rosenblatt, R. Hinde, C. Beer and E. Shaw) Vol. 6, pp. 21–76. Academic Press, New York.

Russek, M., Mogenson, G. J., and Stevenson, J. A. F. (1976). Calorigenic, hyperglycemic and anorexigenic effects of adrenaline and noradrenaline. *Physiol. Behav.* **2**, 429–33.

Saleh, J. W., Yang, M. U., Van Itallie, T. B., and Hashim, S. A. (1979). Ingestive behaviour and composition of weight change during cyproheptadine administration. *Int. J. Obes.* **3**, 213–21.

Samanin, R. and Garattini, S. (1982). Neuropharmacology of feeding. In *Drugs and appetite* (ed. T. Silverstone) pp. 23–39. Academic Press, London.

Samanin, R., Bendotti, C., Bernasconi, S., Borroni, E., and Garattini, S. (1977). Role of brain monoamines in the anorectic activity of mazindol and d-amphetamine in the rat. *Eur. J. Pharmac.* **43**, 117–24.

—— —— —— and Pataccini, R. (1978). Differential role of brain monoamines in the activity of anorectic drugs. In *Central mechanisms of anorectic drugs* (eds. S. Garattini and R. Samanin) pp. 233–42. Raven Press, New York.

—— Caccia, S., Bendotti, C., Borsini, F., Borroni, E., Invernizzi, R., Pataccini, R., and Mennini, T. (1980). Further studies on the mechanism of serotonin-dependent anorexia in rats. *Psychopharmacology* **68**, 99–104.

—— Ghezzi, D., Valzelli, L. and Garattini, S. (1972). The effects of selective lesioning of brain serotonin or catecholamine containing neurones on the anorectic activity of fenfluramine and amphetamine. *Eur. J. Pharmac.* **19**, 318–22.

Sanghvi, S., Singer, G., Friedman, E., and Gershon, S. (1975). Anorexigenic effects of d-amphetamine and *L*-Dopa in the rat. *Pharmac. Biochem. Behav.* **3**, 81–6.

Sclafani, A. (1978). Dietary Obesity. In *Recent advances in obesity research* (ed. G. Bray) Vol. 2, pp. 123–32. Newman, London.

—— and Springer, D. (1976). Dietary obesity in adult rats. Similarities to hypothalamic and human obesity syndromes. *Physiol. Behav.* **17**, 461–71.

Schally, A. V., Redding, T. W., Lucien, H. W., and Meyer, J. (1967). Entergastrone inhibits eating by fasted mice. *Science* **157**, 210–11.

Silverstone, J. T. (1972). The anorectic effects of a long acting preparation of phentermine (Duromine). *Psychopharmacologia* **25**, 315–20.

—— (1982). Drugs and appetite. Academic Press, London.

—— and Fincham, J. (1978). Experimental techniques for the measurement of hunger and food intake in man for use in the evaluation of anorexic drugs. In *Central mechanisms of anorexic drugs* (eds. Garattini, S. and Samanin, R.) pp. 375–82. Raven Press, New York.

—— and Kyriakides, M. (1982). Clinical pharmacology of appetite. In *Drugs and appetite* (ed. T. Silverstone) pp. 93–123. Academic Press, London.

—— and Stunkard, A. (1968). The anorectic effect of dexamphetamine sulphate. *Br. J. Pharmac. Chemother,* **33**, 513–22.

Sofia, R. D. and Barry, H. (1974). Acute and chronic effects of $\triangle^9$-tetrahydrocannabinol on food intake by rats. *Psychopharmacologia* **39**, 213–22.

Steel, J. M. and Briggs, M. (1972). Withdrawal depression in obese patients after fenfluramine treatment. *Br. Med. J.* **3**, 26–7.

Stricker, E. M. and Zigmond, M. J. (1976). Recovery of function after damage to central catecholamine-containing neurones: a neurochemical model for the lateral hypothalamic syndrome. In *Progress in physiological psychology* (eds. J. M. Sprague and A. N. Epstein) Vol. 6, pp. 121–88. Academic Press, New York.

Sugrue, M. F., Goodlet, I., and McIndewar, I. (1975). Failure of depletion of rat brain 5-hydroxytryptamine to alter fenfluramine-induced anorexia. *J. Pharm. Pharmac.* **27**, 950–3.

Sullivan, A. C. and Comai, K. (1978). Pharmacolgical treatment of obesity. *Int. J. Obes.* **2**, 167–89.

—— —— and Triscari, J. (1981*a*). Novel antiobesity agents whose primary site is the gastrointestinal tract. In *Recent advances in obesity research* (eds. P. Bjorntorp, M. Cairella, and A. Howard) Vol. III. Libby, London.

—— Guthrie, R. W., and Triscari, J. (1981*b*). Chlorocitric acid, a novel anorectic agent with a peripheral mode of action. In *Anorectic agents, mechanisms of action and of tolerance* (eds. S. Garattini) pp. 143–58. Raven Press, New York.

—— Nauss-Karol, C., and Cheng, L. (1983*a*). Pharmacological treatment of

obesity I. In *Obesity* (ed. M. R. C. Greenwood) pp. 123–37. Churchill Livingstone. New York.
—— —— —— (1983*b*). Pharmacological treatment of obesity II. In *Obesity* (ed. M. R. C. Greenwood) pp. 139–58. Churchill Livingstone, New York.
—— Triscari, J., Hamilton, J. G., and Miller, O. N. (1974). Effect of (–) -hydroxycitrate upon the accumulation of lipid in the rat: II Appetite. *Lipids* **9**, 129–34.
Tordoff, M. G., Hopfenbeck, J., Butcher, L. L., and Novin, D. (1981). A Peripheral locus for amphetamine anorexia. Unpublished manuscript.
—— Novin, D., and Russek, M. (1980). Truncal vagotomy attenuates amphetamine anorexia. *Soc. Neurosci. Abstr.* **6**, 527.
Trenchard, E., and Silverstone, J. T. (1983). Naloxone reduces the food intake of human volunteers. *Appetite* **4**, 43–50.
Tretter, J. R. and Leibowitz, S. F. (1980). Specific increase in carbohydrate consumption after norepinephine (NE) injection into the paraventricular nucleus (PVN). *Proceedings of the Society for Neuroscience 10th Annual Meeting*, Cincinnati, Nov 9–14, Abstracts Vol. 6, p. 532.
Trygstad, O., Foss, I., Edminson, P. D., Johansen, J. H., and Reichelt, K. L. (1978). Humoral control of appetite: a urinary anorexigenic peptide. Chromatographic patterns of urinary peptides in anorexia nervosa. *Acta Endocrinol.* **89**, 196–208.
Van Itallie, T. B., Gale, S. K., and Kissileff, H. R. (1978). Control of food intake in the regulation of depot fat: an overview. In *Diabetes, obesity and vascular disease* (eds. H. M. Katzen and R. J. Mahler). Wiley, New York.
Vogel, R. A., Cooper, B. R., Barlow, T. S., Prange, A. J., Mueller, R. A., and Breese, G. R. (1979). Effects of thyrotropin-releasing hormone on locomotor activity, operant performance and ingestive behaviour. *J. Pharm. Exp. Ther.* **208**, 161–8.
Wade, G. (1975). Some effects of ovarian hormones on food intake and body weight in female rats. *J. Comp. Physiol. Psychol.* **88**, 183–93.
Wiepkema, P. R. (1971*a*). Positive feedbacks at work during feeding. *Behaviour* **39**, 266–73.
—— (1971*b*). Behavioural factors in the regulation of food intake. *Proc. Nutrit. Soc.* **30**, 142–9.
Wolgin, D. L., Cytawa, J., and Teitelbaum, P. (1976). The role of activation in the regulation of food intake. In *Hunger: basic mechanisms and clinical implications* (eds. D. Novin, W. Wycwicka, and G. Bray) pp. 179–91. Raven Press, New York.
Wooley, S. C. and Wooley, O. W. (1973). Salivation to the thought and sight of food: a new measure of appetite. *Psychosomat. Med.* **35**, 136–41.
Wooley, O. W., Wooley, S. C., and Williams, B. S. (1976). Salivation as a measure of appetite: studies on the anorectic effects of calories and amphetamine In *Hunger: basic mechanisms and clinical implications*. (eds. D. Novin, W. Wyrwicka, and G. A. Bray) pp. 421–29. Raven Press, New York.
—— —— Williams, B. J., and Nurre, D. (1977). Differential effects of amphetamine and fenfluramine on appetite for palatable food in humans. *Int. J. Obes.* **1**, 293–300.
Wurtman, J. J. and Wurtman, R. J. (1977). Fenfluramine and fluoxetine spare protein consumption while suppressing caloric intake by rats. *Science* **198**, 1178–80.
—— —— (1979*a*). Fenfluramine and other serotoninergic drugs depress food intake and carbohydrate consumption while sparing protein consumption. *Current Medical Research and Opinion* Suppl. 1, 28–33.
—— —— (1979*b*). Drugs that enhance central serotoninergic transmission diminish elective carbohydrate consumption by rats. *Life Sci.* **24**, 895–904.

—— (1981). Suppression of carbohydrate consumption as snacks and at meal-time by DL-fenfluramine or tryptophan. In *Anorectic agents: mechanisms of action and tolerance* (eds. S. Garattini and R. Samanin) pp. 169–182. Raven Press, New York.

—— —— Growdon, J. H., Henry, P., Lipscomb, A., and Zeisel, S. H. (1982). Carbohydrate craving in obese people: suppression by treatments affecting serotoninergic transmission. *Int. J. Eat. Dis.* **1**, 2–15.

5

Satietin, a potent and selective endogenous anorectic glycoprotein

J. KNOLL

INTRODUCTION

The recurring desire for food before regular meals, the increased craving for food when an unusual delay between meals occurs, and the rapid abolition of hunger by eating are all common experiences which point to an influence of the gastrointestinal tract on the regulation of hunger and satiety.

On the other hand, the classical demonstration of Hetherington and Ranson in 1940 that lesion of the ventromedial hypothalamus (VMH) in rats leads to hyperphagia, and the brilliant experiments of Anand and Brobeck in 1951, proving that the lesioning of the lateral hypothalamus (LH) in rats and in cats results in hypophagia, led to the promulgation of the theory that feeding is regulated by a 'feeding centre' in the LH and a 'satiety centre' in the VMH (Stellar 1954). The situation is, however, much more complicated. Serious disturbances of food intake follow lesions in the globus pallidus (Morgane 1961), in the substantia nigra (Ungerstedt 1971) and in the n. amygdalae (Fonberg 1974).

The nature of the interaction of these different brain areas, as well as the involvement of catecholaminergic, serotonergic and other transmitter pathways in the regulation of feeding are still matters of extensive investigations (see Chapter 00). But, be that as it may, detailed physiological, pharmacological and behavioural studies on feeding (Blundell and McArthur 1979; Hoebel and Novin 1982; Novin, Wyrwiczka, and Bray 1976; Silverstone 1982; Wayner and Oomura 1975) are consistent with the view that what is primarily regulated in the brain is satiety.

There is, however, an essential gap, a missing link of basic importance, in our current knowledge about the regulation of food intake. We have no satisfactory explanation of the fact that, whereas, the craving for food gradually and relatively slowly, increases during starvation, e.g. days of food deprivation are needed to reach the peak of the hunger drive in a rat, the desire to ingest food is rapidly abolished during feeding.

Earlier detailed physiological studies (Knoll 1969) of the hunger-drive-induced behaviour in rats prompted me to assume that a highly potent and highly selective anorectic substance may be activated or liberated in the blood during eating which terminates the hunger drive. Figure 5.1 shows the phenomenon which enforced this working hypothesis.

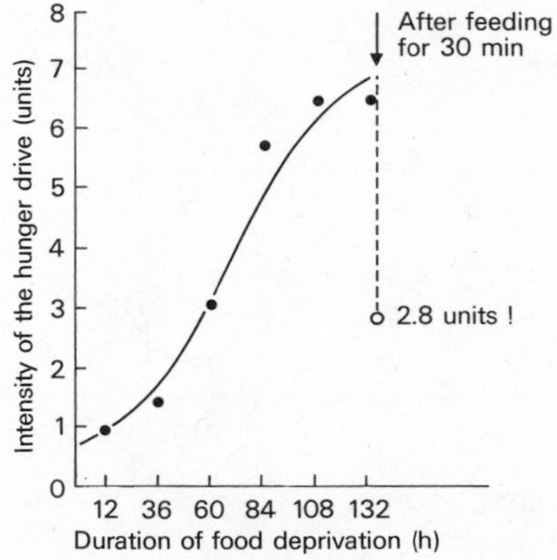

FIG. 5.1 The gradual increase of the intensity of the hunger drive in rats during 132 h starvation and its extinction within a short period of feeding. The intensity of the hunger drive was expressed in units according to a method based on the assessement of the spontaneous motility of strarving rats in an open field (Knoll 1969, Ch. II). The experiment was performed with a group of 20 rats weighing 200–250 g before starvation and supplied with water *ad libitum*. Chow pellets were offered for 30 min after 132 h starvation.

We studied the hunger drive in the rat, and elaborated, by measuring the spontaneous motility of the starving animal in an open field, a method for expressing the intensity of the drive in units from zero to ten (Knoll 1969, Chapter II). Figure 5.1 shows that the hunger-induced specific activation of the central nervous system, gradually increases during starvation in rats deprived of food for 132 h. Feeding for 30 min was, however, sufficient in this series of experiments to decrease the intensity of the drive from 6.2 to 2.8 units. As feeding reduces the desire for food so rapidly, the hypothesis was formulated that food consumption initiates the liberation or activation of a blood-borne, rate-limiting satiety signal which by stimulating selectively the satiety centre terminates the starving-induced activation of the central nervous system.

The development of gel-chromatographic techniques for the separation of endogenous substances allowed us to check the validity of this hypothesis. Experiments were started in October 1976. Human serum was filtered through an Amicon UM 10 membrane. The ultrafiltrate was evaporated and chromatographed on a Sephadex G-15 column, using 0.9 per cent NaCl as eluent in order to make the measurement of the anorectic acitivity in the fractions easier. As the phenomenon demonstrated in Fig. 5.1 can only be explained by a highly potent blood-borne inhibitor of food intake, rats deprived of food for 96 h were used for testing the anorectic activity of the fractions. We collected 10 ml fractions and injected 5 ml/kg intravenously (i.v.) using rats weighing about 200 g after starvation. Thus, each fraction could be checked in 8–10 rats deprived of food for 96 h. Food was offered 1 h after the injection of the test material. Food intake during the first hour and the 24 h consumption was measured. A previously unknown substance in human serum capable of suppressing food intake in rats deprived of food for 96 h, which, I named satietin, was successfully detected in May 1977 (Knoll 1978; Knoll 1979*a*).

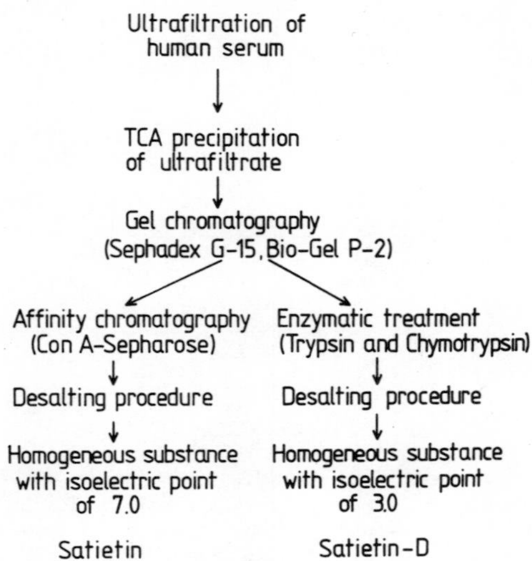

FIG. 5.2 Flow sheet of the isolation procedure of satietin and satietin-D.

CHEMICAL NATURE OF SATIETIN AND SATIETIN-D ISOLATED FROM HUMAN SERUM

Figure 5.2 shows the flow sheet of the isolation of human serum satietin, and one of its enzymatic digestion products, satietin–D.

Satietin, isolated by means of gel-chromatography and affinity chromatography from human serum, is a glycoprotein with a molecular weight of 64–69000 daltons. Its isoelectric point is 7.0. It contains 14–15 per cent amino acids and 70–75 per cent carbohydrates. Its biological activity survives digestion with proteases (trypsin, chymotrypsin and carboxypeptides) and boiling (Knoll 1979*a*, 1980, 1982*b*, 1982*c*; Nagy, Kalasz and Knoll 1982, 1983).

Satietin–D is a homogeneous product, isolated from human serum by means of gel-chromatography and enzymatic digestion. Its molecular weight is 41–43 000 dalton, but it has an isoelectric point of 3.0, a protein content of 20–22 per cent and a carbohydrate content of 55–60 per cent. The biological activity survives boiling. The most important characteristics of the two homogeneous glycoproteins are presented in Table 5.1

The homogeneity of satietin and satietin–D was proved by polyacrylamide gel-electrophoresis in the presence of sodium dodecyl sulfate (SDS), gradient gel electrophoresis in slab gel and analytical isoelectric focusing.

Satietin and satietin–D are identically stained with ninhydrin and periodic acid-Schiff reagents indicating the glycoprotein nature of these substances.

Amino acids in satietin and satietin–D were determined after acid hydrolysis by microbiuret and by means of an amino-acid analyser.

The amino-acid composition in satietin and satietin–D, respectively, expressed in per cent, was found as follows:

	Satietin (%)	Satietin-D (%)
Asp	1.21	2.83
Thr	0.66	1.61
Ser	0.71	1.25
Glu	1.87	3.60
Pro	0.45	1.36
Gly	0.42	0.87
Ala	1.58	1.18
Cys	0.13	0.16
Val	0.55	0.93
Met	0.12	0.17
Ile	0.26	0.76
Leu	0.90	1.25
Tyr	0.31	1.42
Phe	0.44	0.77
Lys	3.79	3.67
His	0.15	0.22
Arg	0.49	0.67
Total	14.04	22.72

TABLE 5.1 *Characteristics of satietin and satietin–D samples isolated from human serum*

	Satietin	Satietin–D
Molecular weight (daltons)	64–69000	41–43000
Isoelectric point	7.0	3.0
Protein content (%)	14–15	20–22
Carbohydrate content (%)	70–75	55–60
Water content (%)	5–10	5–10
Anorectic activity in rats (intra-cerebroventricular ID_{50}; μg/rat)	10–20	10–20

In both substances mannose, galactose, fucose, glucose and glucosamine are the detected carbohydrate components.

Freeze-dried satietin and satietin–D are white, easily water-soluble, and stable amorphous powders.

SATIETIN IN THE SERUM OF DIFFERENT MAMMALS AND IN POULTRY BLOOD

Satietin was found to be present in the serum of all the species selected for examination (mouse, different strains of rats (Wistar, Long Evans, CFY, SHR), guinea pig, and rabbit) (Knoll 1979*b*, 1980).

Satietin was also detected in bovine and horse sera, (ungulates), and in the serum of the cat and dog (carnivora) (Knoll 1979*b*, 1979–80, 1980).

We were able to obtain a few litres of goose serum and thus we selected this species to check the presence of satietin in avian blood. We found an anorectic substance in goose serum which was gel-chromatographically indistinguishable from the active substance prepared from the sera of mammals (Knoll 1982*c*).

The isolation of satietins detected in horse, cattle, and goose sera is in progress. According to preliminary observations these satietins seem to be closely related glycoproteins, but not identical with each other or with human serum satietin.

ANORECTIC EFFECT OF SATIETIN

Satietin proved to be a highly potent anorectic substance which inhibited food intake dose-dependently in rats deprived of food for 96 h. Table 5.2 shows the dose-dependent and long-lasting suppression of food intake in hungry rats by satietin isolated from human serum.

A biological assay for measuring satietin acitivity in units was developed. The unit is equivalent to the anorexogenic activity of the amount of a satietin sample

TABLE 5.2 *An example of the bioassay of a highly purified satietin sample by measuring its dose-dependent anorectic effect in rats deprived of food for 96 h*

Treatment		Food intake mean ± S.E.M.		
	μg/rat	g/1 h	g/24 h	
Saline	20	7.58 ±0.63	23.92 ±0.99	
	μg/rat			Satietin activity expressed in units
Satietin	2.5	5.58 ±0.79	19.02 ±0.87	0.25
	5	4.90* ±0.68	13.18 ±0.69	0.50
	7.5	3.88* ±0.92	12.15* ±1.19	0.75
	10	1.88† ±0.41	9.82† ±0.87	1.00
	20	0.62† ±0.28	6.51† ±0.94	2.00
	40	0.78† ±0.34	3.22† ±1.29	4.00

Rats were deprived of food for 96 h. Water was supplied *ad libitum*. The satietin sample was prepared from human serum. Satietin dissolved in saline was injected in a volume of 20 μl/rat intracerebroventricularly into the lateral ventricle 1 h before feeding. Each dose was tested on 16 rats. The consumption of chow pellets (g) during the first hour of feeding (g/h) and during 24 h (g/24 h) is shown in the Table.

Statistics: Student's t-test for two means $^*p < 0.05$; $^†p < 0.001$.

which, when given intracerebroventricularly, decreases the chow pellet consumption of rats deprived of food for 96 h during the first day of feeding from 24.04 ±0.76 g to 10 g. With higher intracerebroventricular doses of satietin (2–3 units/rat) the 24 h consumption of the fasting rats can be reduced to 3–5 g, and the animals begin to eat on the second day of feeding only. The satietin sample, the effect of which is shown in Table 5.2 contained according to the definition of the arbitrary unit, 100 units/mg activity.

Satietin samples exert a drug-dependent anorectic effect in rats, both on intravenous and on subcutaneous administration. Isolated satietin and satietin-D were found to be equally potent anorectic substances which inhibit food intake in a dose-dependent manner. The intracerebroventricular administration of 10 μg satietin suppressed food intake in rats during the first day of feeding after deprivation of food for 96 h to half of the amount eaten by untreated controls (ID_{50}). The onset of the effect can be detected within 30 min, the peak being reached within an hour. The effect lasts 24–30 h. Following intravenous administration the ID_{50} of satietin and satietin-D was found to be 0.5–0.75 mg/kg in rats deprived of food for 96 h. The peak effect was reached within 1 h, and lasted over 24 h.

Most of the biological information on satietin stems from intracerebroventricular administration of the substance. The difficulties involved in the isolation of satietin have so far restricted the parenteral administration of highly purified material, and permit virtually only the intracerebroventricular application of pure satietin and satietin-D. For the same reasons the long-term anorectic efficiency of satietin also remains to be investigated.

Table 5.3 shows the unchanged efficiency of two consecutive doses of homogeneous satietin in a group of normally-fed rats. The effect of the first dose of satietin lasted about 48 h and the food consumption returned to normal on the third day after the intracerebroventricular injection of satietin. There was no sign of rebound. The course of effect of the second dose of satietin was identical with that of the first dose. Food consumption was severely depressed during 48 h after the injection of satietin, but there was no significant difference in the third day that consumption was to be observed. Food consumption remained normal on the consecutive days.

These experiments induce the hope that, in spite of the long-lasting anorectic

TABLE 5.3 *The unchanged efficiency of two consecutive doses of satietin and the lack of rebound in normally fed rats*

Days	1	2	3	4	5	6	7	8	9	10	11	12
Daily food consumption (g)	26.2 ↑ satietin	6.89*	14.8*	27.4	25.2	24.2 ↑ satietin	4.5*	13.2*	21.8	26.6	29.2	27.4

*significant $p < 0.001$.

$n = 12$; 40 μg satietin was injected intracerebroventricularly (right and left side).

effect of satietin, the continuous administration of the substance will not change the onset and offset of its anorectic effect and it will not interfere with the normal endogenous regulation of food intake. However, confirmation or otherwise of this view depends on the availability of an abundant supply of satietin, in order to perform the necessary long-term experiments.

FOOD-INTAKE SUPPRESSING EFFECT OF SATIETIN IN COMPARISON TO OTHER ENDOGENOUS ANORECTIC SUBSTANCES

During the last decade a number of endogenous substances, mainly peptides, have been reported to decrease food intake in animals. Table 5.4 shows the variety of endogenous substances, mostly known peptide hormones, described as having a food-intake suppressing effect. The list is, at first sight, perplexing.

TABLE 5.4 *List of endogenous anorectic substances in chronological order of recognition of their effect of feeding (satietin is omitted)*

1957	Glucagon (Schulman *et al.* 1957)
1964	PGE_2 and $PGF_{2\alpha}$ (Horton 1964)
1967	Enterogastrone (Schally *et al.* 1967)
1973	Cholecystokinin (Gibbs *et al.* 1973, 1976)
1977	Thyrotropin releasing hormone (Vijayan and McCann 1977)
	Beta-endorphin → Naloxone (Grandison and Guidotti 1977)
	Pancreatic polypeptide (Malaisse-Lagae *et al.* 1977)
1978	*p*Glu-His-GlyOH (Reichelt *et al.* 1978)
1979	Somatostatin (Lotter *et al.* 1981)
	Bombesin (Gibbs *et al.* 1979)
	Calcitonin (Freed *et al.* 1979)
	Insulin (Woods *et al.* 1981)
1981	Vasoactive intestinal peptide (Woods *et al.* 1981)
1982	Corticotropin releasing hormone (Morley and Levine 1982)
	2-Deoxytetronate (Oomura *et al.* 1982)
	Neurotensin (Hoebel *et al.* 1982)

If we look at the chronological order of the recognition of their effects on feeding, we can see that 12 out of the 16 items on the list were described during the last seven years, reflecting the increased interest in this field. The variety of the listed endogenous substances is no longer perplexing if we group them as shown in Table 5.5.

Of the 16 substances reported to inhibit food intake, eight (enterogastron, cholecystokinin, bombesin, vasoactive intestinal peptide, glucagon, insulin, pancreatic polypeptide, and 2-deoxy-tetronate) are related to the alimentary tract or metabolism. None of them is a supressor of feeding in a rat deprived of food for 96 h, i.e. at the peak of the hunger drive. The same is true for the second group: thyrotropin-releasing hormone, corticotropin-releasing hormone, neurotensin, which increases the secretion of ACTH and glucagon, and somatostatin, which inhibits the release of somatotropin. All members of this group of

TABLE 5.5 *Grouping of endogenous substances reported to possess an anorectic effect (Satietin is omitted)*

Related to alimentary tract or metabolism	Enterogastrone CCK Bombesin *Vasoactive intestinal polypeptide* Glucagon Insulin Pancreatic polypeptide 2-Deoxytetronate
Hypothalamic hormones or structural relatives	TRH (pGlu-His-Pro-NH_2) CRF Somatostatin *Neurotensin* pGlu-His-Gly-OH
Non-specific systems	Prostaglandins Enkephalins (Naloxone)
Thyroid hormone	Calcitonin

hormones are related to the hypothalamus, thought to be a main site of the central regulation of feeding.

There are also two non-specific elements in the list: prostaglandins and endogenous opioids.

The prostaglandins, which have some influence on almost all functions ever investigated, are only of slight theoretical interest from the point of view of food intake regulation.

The enkephalins have no selectivity either; they seem to influence too many brain functions. However, the 'pure opiate' antagonists, which are safe, well-tolerated compounds, have been shown to have an anorectic effect (see Chapter 00). The development in the near future of a special new family of antiobesity agents, acting via opiate receptors, seems to be a possibility.

Of the list in Table 5.4 only calcitonin is as potent as satietin in suppressing food intake in rats deprived of food for 96 h; all the other endogenous substances reported to have anorectic effects are ineffective in animals with similar intensive hunger drives. Calcitonin has a specific hormonal action: it inhibits bone resorption by altering osteoclastic and osteocytic activity. Considering its high potency in influencing calcium metabolism which is in harmony with the very low physiological concentrations (70–120 pcg/ml) of this hormone, it inhibits food intake in relatively high amounts only. The reduction of feeding in rats by calcitonin is probably secondary to its inhibition of calcium uptake into hypothalamic neurones (Levine and Morley 1981). Even if calcitonin lacks the selectivity of satietin, its potential role as a blood-borne satiety signal requires consideration. If we take into account that the satietin concentration in human serum is over 2 μg/ml, the difference on a weight basis is about 20 000 and on a molecular basis (50 000 versus 3600) about 1500. The intracerebroventricular efficacy of calcitonin on a weight basis is about 10 times higher than that of satietin, but satietin is a 13.88 times bigger molecule. Thus, the two peptides can be taken to be roughly equally active as anorectic agents.

Comparing the anorectic activities and the blood concentrations of satietin and calcitonin it is difficult to believe that calcitonin plays a prominent part in the physiological regulation of feeding as a blood-borne satiety signal when we have in our blood an anorexogenic glycoprotein, satietin, in a molar concentration, at least 1500 times higher which is at least as potent as calcitonin in suppressing food intake, and is in addition much more selective. Thus, calcitonin's effect as a blood-signal would appear negligible beside that of satietin. However, as calcitonin is also released in the brain, it might be involved in the physiological regulation of feeding as a locally-acting hormone in the hypothalamus.

FOOD-INTAKE SUPPRESSING EFFECT OF SATIETIN IN COMPARISON TO ANORECTIC DRUGS

Phenylisopropylamine (amphetamine) was the first drug used as an antiobesity agent. In the past decade there have been a number of studies designed to analyze which neurotransmitter functions are involved in the regulation of hunger and satiety (see Chapter 4 for further discussion of this).

The anorectic effect of satietin is different from that of any of the standard anorectic drugs. The first important difference is in the duration of the effect. Single doses of amphetamine, fenfluramine, and mazindol have no significant influence on the 24 h consumption in rats after 96 h of food deprivation because their anorectic effect lasts only 4–5 hours.

The anorectic effect of amphetamine and mazindol, that act via the release of catecholamines, is considerably weakened in rats pretreated with α-MT, which inhibits the synthesis of the catecholamines (Knoll 1979*a*, 1980, 1982*a*), whereas the effect of satietin remains unaltered in α-MT pretreated animals.

The anorectic effect of fenfluramine was prevented by lesioning the median and dorsal raphe (Knoll and Knoll 1982), whereas the effect of satietin remained unaltered in the lesioned rats.

The satietin effect remained unchanged also in substantia-nigra-lesioned animals, but bilateral lesioning of the ventromedial hypothalamus seemed to weaken the efficiency of satietin (Knoll and Knoll 1982).

The time course of the effect of satietin in comparison to that of amphetamine, fenfluramine and calcitonin was studied in an eatometer. Blundell found differences in the anorectic effect of amphetamine and fenfluramine by analysing the microstructure of eating, and concluded that amphetamine which activates catecholaminergic transmission acts as a suppressor of the onset of eating, whereas fenfluramine facilitates satiety (Blundell and Latham 1982).

By measuring the time from the beginning of the experiment to the consumption of the first bite in rats following 24 h starvation, we were able to corroborate the finding of Blundell. Amphetamine (1 mg/kg, s.c.) significantly extended the latency time of the beginning of eating, whereas fenfluramine (3 mg/kg, s.c.), satietin (4 mg/kg, s.c.) and calcitonin (10 MRC units/kg, s.c.) all injected 30 min before feeding, did not change the onset of eating (Knoll 1982*b*, 1982*c*; G. Sandor and J. Knoll, to be published).

Figure 5.3 shows the typical curves for fenfluramine and amphetamine. The two doses were selected to be equipotent in inhibiting the total consumption of food during 1 h of feeding in rats deprived of food for 24 h. It is evident from this figure that fenfluramine is less active in inhibiting food intake in the first 15 min than amphetamine. The amphetamine-treated rats eat equally small amounts of food during the four consecutive quarters of the 1 h feeding period, fenfluramine blocks eating completely during the second half of the feeding period.

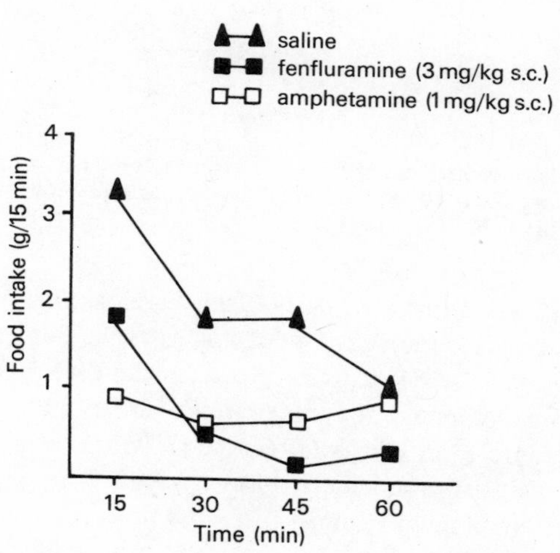

FIG. 5.3 The time course of the effect of fenfluramine and amphetamine in comparison to saline-treated rats deprived of food for 24 h. Measurements were performed in an eatometer. Both substances were administered subcutaneously 30 min before feeding.

Figure 5.4 shows that both satietin and calcitonin act similarly to fenfluramine; they are not very active in inhibiting food consumption during the first 15 min period of feeding, but later completely suppress feeding. For sake of comparison, the time structure of food intake under the influence of amphetamine is also shown in the Fig. 5.4.

These experiments speak in favour of the assumption that satietin acts by facilitating satiety. In this respect fenfluramine acts similarly, but there is an important difference in the mechanism of action between the two substances, as the anorectic effect of fenfluramine is prevented by the lesioning of the raphe system, whereas that of satietin remains unaltered in the lesioned animals.

To illustrate the anorectic potency of satietin in comparison to anorectic drugs, the relative potency of satietin to fenfluramine in rats deprived of food for 96 h was evaluated. In comparison to satietin, fenfluramine was short-acting. Only the consumptions in the first hour are comparable. The compounds were

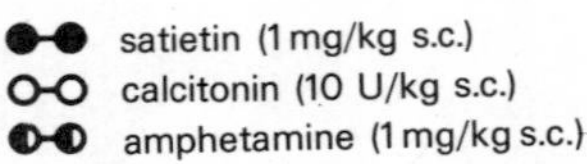

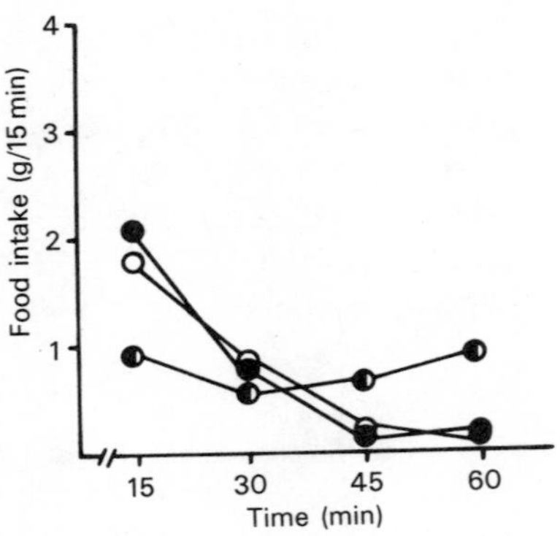

FIG. 5.4 The time course of the effect of satietin in comparison to that of calcitonin and amphetamine in rats deprived of food for 24 h. Substances were administered subcutaneously 30 min before feeding.

administered intracerebroventricularly 1 h before feeding. About 50 μg fenfluramine was found to be equivalent to 20 μg satietin. Considering the molecular weights (50 000 versus 181) satietin proved to be 690 times more potent than fenfluramine in suppressing food intake in the hungry rats, when given intracerebroventricularly. By comparing the relative activities at intravenous administration we found that, on a molecular basis, satietin was at least 500 times more potent than fenfluramine in inhibiting the first hour consumption of food in rats deprived of food for 96 h.

SELECTIVITY OF THE ANORECTIC EFFECT OF SATIETIN

The anorectic drugs in medicinal practice act by releasing either catecholamines or serotonin, and because any of the biogenic amines released by an anorectic drug is involved in different functions in the brain and in the periphery, none of the anorectics inhibit food intake selectively. The lack of selectivity leads to a number of side effects and anorectic drugs are mainly considered to be short-term adjuvants in a more complex therapy of obesity, including calorie restriction, appropriate exercise and psychological support.

The potential clinical value as anorectic agents, of any of the other endogenous substances listed in Table 5.4, is doubtful, as they all have characteristic physiological activities exerted in much lower than anorectic concentrations.

Satietin seems to be, at present, the only known endogenous substance which does not exert any other noticeable central or peripheral effect in the anorectic dose range.

Table 5.6 summarizes the effects of equi-anorectic doses of satietin, calcitonin, amphetamine and fenfluramine on the behaviour of rats in a battery

TABLE 5.6 *Comparison of the effect of satietin, calcitonin, amphetamine and fenfluramine on the behaviour of rats in a battery of tests*

	Number of rats	Satietin i.c.v.	Calcitonin i.c.v.	Amphetamine i.v.	Fenfluramine i.c.v.	Method
Locomotor activity	10	None	None	Strong facilitation	Strong inhibition	Open field
One-way avoidance	10	None	None	Strong facilitation	Strong inhibition	Modified jumping test (Knoll and Knoll 1964)
Two-way avoidance	12	None	None	Facilitation	Inhibition	Shuttle-box
One way conditioning	10	None	Strong inhibition	Strong facilitation	Strong inhibition	Screening test I (Knoll B. *et al.* 1974).
Consolidated conditioned reflex	6	None	Inhibition	None	Strong inhibition	Jumping test (Knoll and Knoll 1958, 1959)
Male copulatory behaviour	13	None	–	Facilitation	Inhibition	Beach 1944

All compounds were administered in the dose equi-anorectic with satietin (usually 1–2 units) either intracerebroventricularly (i.c.v.) or intravenously (i.v.) to group of rats. Satietin was isolated from human serum. For methodological and other details see Knoll and Knoll 1982 and Yen, Dallo, and Knoll 1982.

of tests. Fenfluramine, in an anorectic dose, was found to be a strong inhibitor in all the tests studied. It decreased locomotion in the open field, strongly interfered with unconditioned avoidance reactions, inhibited the development of conditioned reflexes, blocked learning and retention in one-way and two-way avoidance systems, inhibited the recall of a previously firmly-developed conditioned response, and completely inhibited copulatory behaviour in male rats. In contrast to fenfluramine, amphetamine was stimulatory in the tests and facilitated performances. The anorectic dose of calcitonin inhibited the acquisition of a conditioned reflex in a one-way avoidance system and blocked the recall of a firmly established conditioned response; it was without effect in the three other tests.

Satietin had no effect on any of the tests. Table 5.7 compares the effects of satietin, calcitonin, amphetamine and fenfluramine on the metabolic rate, body temperature and blood pressure. Again, none of these parameters were influenced by a dose of satietin which blocked food intake in the rat completely.

That satietin is highly selective in suppressing food consumption is further supported by the finding that the intracerebroventricular administration of 1–2 units of satietin into the lateral ventricle, which exerts a strong anorectic effect, has no effect on the water intake of rats. Figure 5.5. shows an example. Groups of rats were deprived of water for 23 h and injected intracerebroventricularly 1 h before water supply with satietin (0.25, 0.5, 1, and 2 U) and calcitonin (0.125, 0.25, 0.5, and 1 U), respectively. The control group was treated with saline, 20 μl/animal. Water was given at the end of the 23 h deprivation period and its consumption was measured during 4 h. The animals were starving during the experiment. None of the satietin doses caused significant change compared to

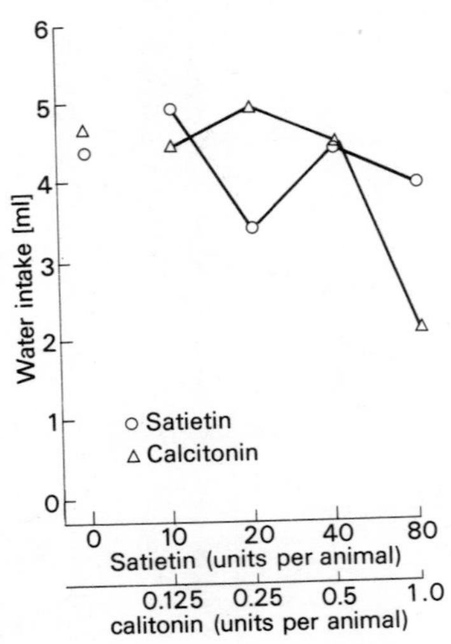

FIG. 5.5 Effect of satietin and calcitonin on the water intake of rats in the first 4 h after 23 h water deprivation. 7 rats are in each group. Satietin and calcitonin were injected into the lateral ventricle in the 22nd hour of water deprivation.

TABLE 5.7 *Comparison of the effect of satietin, calcitonin, amphetamine and fenfluramine on metabolic rate, body temperature and blood pressure in the rat*

	Satietin	Calcitonin	Amphetamine	Fenfluramine	Method
Metabolic rate	None	Significant increase (38%)	Significant increase (97%)	None	Issekutz and Issekutz 1942
Body temperature	None	Slight, statistically insignificant elevation	Slight, statistically insignificant elevation	None	Continuous measurement of rectal temperature
Blood pressure	None	None	Significant increase	Slight decrease	Via the carotid artery

All compounds were administered intracerebroventricularly in doses equi-anorectic with the dose of satietin. Satietin prepared from human serum was used in the experiments. For methodological and other details see Timár and Knoll 1982.

the controls, while 1 MRC unit of calcitonin decreased water intake significantly ($p < 0.01$).

Satietin did not alter the water intake of rats deprived of food for 96 h and supplied with water *ad libitum*, as shown in Table 5.8. It can be seen from the data in Table 5.8 that rats fed with standard chow pellets need, because of the dry food, a high amount of water for consuming the pellets. During the 96 h starvation period the water consumption of the rats decreased to their essential need, which is about 6–8 ml daily. In contrast to satietin, amphetamine (5 mg/kg subcutaneously every 5 h) doubled the water intake of rats (12.83±1.92 ml versus 6.45±0.83 ml), under the same experimental circumstances.

We have also demonstrated, by using the 'conditioned aversion' paradigm of Garcia, Hawkins, and Kenneth (1974), that satietin, even in higher than anorectic dose, did not elicit aversion for food (Sandor and Knoll 1982), i.e. it is a real anorectic agent.

The intracerebroventricular administration of satietin did not change the levels of glucose, insulin or glucagon in the blood, either in normally fed rats or in rats deprived of food for 96 h. Amphetamine, given intracerebroventricularly, did not change these parameters in normally fed animals but significantly increased the serum insulin level in food-deprived rats (Gyarmati, Földes, Korányi, Knoll, and Knoll 1982).

Amphetamine, which is a potent releaser of noradrenaline exerts a number of peripheral effects and is highly potent on isolated organs possessing noradrenergic transmitter systems. We checked the effect of satietin on the perfused central ear artery of the rabbit, on the pulmonary artery strip of the rabbit and on the isolated nictitating membrane of the cat, in which less than 1 μm/ml amphetamine strongly potentiates the response of the vascular smooth muscle to nerve stimulation. Even very high doses of satietin proved to be completely without activity in these tests (Knoll 1982*b*, 1982*c*).

We tested the effect of satietin in spinal rats by measuring the contraction of

TABLE 5.8 *Ineffectiveness of satietin on the water intake of rats deprived of food for 96 h*

Treatment	Dose	Water intake (ml) ± S.E.M.					
		Before starvation	After starvation (days)				After feeding
			1	2	3	4	
Without treatment	–	37.4 ±1.66	13.75 ±2.0	12.95 ±1.18	8.35 ±0.79	6.45 ±0.83	35.9 ±2.79
Saline*	20 μ/animal	32.28 ±1.08	15.7 ±2.12	10.0 ±0.89	7.0 ±0.63	6.6 ±0.94	38.9 ±1.15
Satietin*	80 μ/animal	33.75 ±1.48	13.05 ±1.31	12.1 ±1.18	9.6 ±0.89	6.2 ±0.88	37.8 ±1.29

* Injected in the morning of the 4th day's starvation.
$n = 20$. Intracerebroventricular administration

the m. tibialis anterior following hind paw stimulation. Serotonin is known to be involved in the spinal reflex and fenfluramine, given intravenously at doses, lower than those causing anorexia, increases the contractions of the m. tibialis anterior by enhancing the activity of the serotonergic link in the reflex. Satietin proved to be completely inactive in this respect when given intravenously in a much higher than anorectic dose (Knoll 1982*b,c*).

The effect of satietin was also investigated on isolated organs (longitudinal muscle strip of the guinea pig ileum, mouse vas deferens and cat splenic strip) which are used for testing opiate agonists and antagonists. Satietin did not exert any effect in these tests and failed to influence the effect of opiate agonists.

PUTATIVE ROLE OF SATIETIN IN THE REGULATION OF FOOD INTAKE

Putting all the data together, the working hypothesis visualized in Fig. 5.6 is put forward.

Considering that satietin is widely distributed in vertebrate species, that its concentration in the blood is high, that its site of effect is in the central nervous system, and that it induces satiety without having any other detectable central or peripheral effect, *it may play the role of a rate-limiting, blood-borne, satiety signal in the negative feed-back of food intake, i.e. serving as the essential chemical link connecting the gastrointestinal tract and the brain in the regulation of feeding.*

The high selectivity of satietin might be explained by the assumption that neurones in the brain with highly specific satietin receptors exist and function as a 'satiety centre'. If, because of the high satietin concentration in the blood, these receptors are saturated with satietin, food intake is completely inhibited, and we have the subjective feeling of fullness, and possibly even an aversion for food. Satietin might be liberated from inactive binding by special signals. Regulator(s) of the liberation of physiologically active satietin may play important roles in satiation and there is space for short term satiety signal(s) originating from the gastrointestinal tract.

That satietin circulates in the blood in an inactive form is strongly supported by the finding that, whereas with the material prepared by the method without the step of trichloroacetic acid precipitation (Knoll 1982*c*), the peak effect was reached 5 h after the intracerebroventricular injection of the sample, the trichloroacetic acid precipitation of albumin before gel chromatographic separation (see Fig. 5.2) yielded preparations with a much shorter (30 min) onset of action.

Digestion of food, and the emptying of the gastrointestinal tract, may lead to a gradual decrease of active satietin in the blood, with a corresponding fall in concentration in the brain; this leads to the inactivation of the 'satiety centre' and to the progressively intesive feeling of hunger as visualized in Fig. 5.6

If satietin were the blood-borne rate-limiting satiety signal in the negative feed-back of food intake this would have important theoretical and practical consequences. Theoretically, it would mean a decisive step forward in the elucidation of the biogrammar of feeding. From a practical point of view, a new

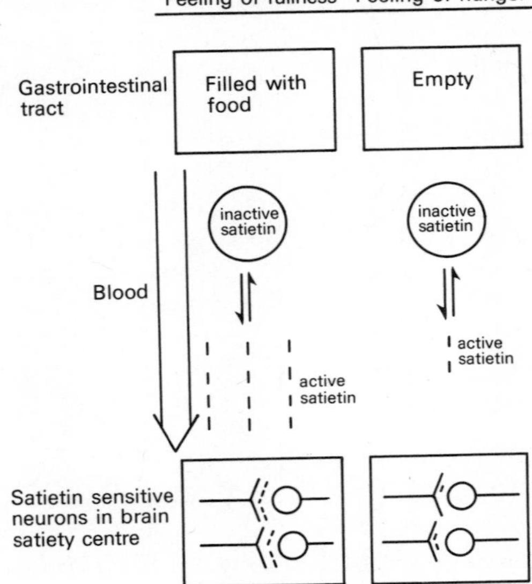

FIG. 5.6 Scheme visualizing the hypothesis that satietin, a blood-borne substance, plays the role of a rate-limiting satiety signal in the negative feed-back of food intake.

opportunity would have presented itself for influencing food intake. According to the hypothesis, hunger is terminated by the liberation of physiologically active satietin in the blood which is brought by eating an appropriate amount of food. As the brain should be satiated by satietin, if we could learn to raise, to a sufficient extent, the blood level of the satiety signal by exogenous manipulation, we could terminate the feeling of hunger by avoiding the natural route: filling the gastrointestinal tract. We may thus have the option to terminate the hunger drive either through the 'natural' method, that is by eating, or artificially by increasing the satietin concentration in the blood exogenously. To evaluate this possibility further is one of the main objects of our research.

References

Anand, B. K. and Brobeck, J. R. (1951). Hypothalamic control of food intake in rats and cats. *Yale Biol. Med.* **24**, 123–40.

Beach, F. A. (1944). Relative effect of androgens upon the mating behaviour of male rats subjected to forebrain injury or castration. *Exp. Zool.* **97**, 249–59.

Blundell, J. E. and Latham, C. J. (1982). Behavioural pharmacology of feeding. In *Drugs and appetite.* (ed. T. Silverstone) pp. 41–80. Academic Press, London.

—— and McArthur, R. A. (1979). *Obesity and its Treatment*, Vol. 1, p. 104. Eden Press, Montreal.

Fonberg. E. (1974). Amygdala functions within the alimentary system. *Acta Neurobiol. Exp.* **34**, 435–66.

Freed, W. J., Perlow, M. J., and Wyatt, R. J. (1979). Calcitonin inhibitory effect on eating in rats. *Science* **206**, 850–52.

Garcia, J., Hankins, W. G., and Kenneth, R. W. (1974). Behavioural regulation of the milieu interne in man and rat. *Science* **185**, 824–31.

Gibbs, J., Fauser, D. J., Rowe, E. A., Rolls, B. J., Rolls, E. T., and Madison, S. P. (1979). Bombesin suppresses feeding in rats. *Nature* **282**, 208–10.

—— Falasco, J. D., and McHugh, P. R. (1976). Cholecystokinin elicits satiety in rats with open gastric fistulas. *Nature* **245**, 323–25.

—— Young, R. C., and Smith, G. P. (1973). Cholecystokinin decreases food intake in rats. *J. Comp. Physiol. Psychol.* **84**, 488–95.

Grossman, S. P. (1960). Eating and drinking elicted by direct adrenergic and cholinergic stimulation of hypothalamus. *Science* **132**, 301–2.

Grandison, S. and Guidotti, A. (1977). Stimulation of food intake by muscimol and beta-endorphin. *Neuropharmacology* **16**, 533–36.

Gyarmati, S., Földes, J., Korányi, L., Knoll, B., and Knoll, J. (1982). A szatietin hatása a szénhidrát anyagcserére. *Orvostudomány* **33**, 547–51.

Hetherington, A. W. and Ranson, S. W. (1940). Hypothalamic lesions and adiposity in the rat. *Anat. Rec.* **78**, 149–72.

Hoebel, B. G. and Novin, D. (eds.) (1982). *The neuronal basis of feeding and reward.* Haer Institute, Brunswick, Massachusetts.

—— Hernandez, L., and McLean, S. (1982). Catecholamines, enkephalin and neurotensin in feeding and reward. In *The neuronal basis of feeding and reward.* (eds. B. G. Hoebel and D. Novin) pp. 465–78, Haer Institute, Massachusetts, Brunswick.

Horton, E. W. (1964). Actions of prostaglandins E_1, E_2 and E_3 on the central nervous system. *Br. J. Pharmac. Chemother.* **22**, 189–92.

Issekutz, B. and Issekutz, B. Jr. (1942). Einfache Apparate zur Messung des Sauerstoff-Verbrauches von Versuchstieren. *Arch. Exp. Pathol. Pharmac.* **199**, 306–11.

Knoll, B. and Knoll, J. (1982). The selectivity of the anorectic effect of satietin. I. The ineffectiveness of satietin in behavioural tests. *Pol. J. Pharmac. Pharm.* **34**, 17–23.

—— Held, K. and Knoll, J. (1974). Rapid screening of drug action on learning and memory. In *Symposium on pharmacology of learning and retention* (ed. B. Knoll) pp. 43–47. Academiai Kiado, Budapest.

Knoll, J. (1969). *Theory of active reflexes.* Akademiai Kiado, Budapest.

—— (1978). New aspects in the chemoregulation of food intake. *Neurosci. Letts.* Suppl. 1. S55.

—— (1979*a*). Satietin: a highly potent anorexogenic substance in human serum. *Physiol. Behav.* **23**, 497–502.

—— (1979*b*). Az extrém tápláltsági állapotok gyógyszertani vonatkozásai. In *A Korányi Társaság Tudományos Ulései. XVI. Az extrém tápláltsági állapotok.* (eds. S. Eckhardt, Gyenes, G.) pp. 29–44. Akademiai Kiadó, Budapest.

—— (1979–1980). Szatietin: a táplálékfelvételt szelektiven gátló anyag az emberi vefben. *Orvostudomány* **30–31**, 351–83.

—— (1980). Highly selective peptide chalones in human serum. In *Modulation of neurochemical transmission* (ed E. S. Vizi) pp. 97–125. Akadémiai Kiadó-Pergamon Press, Budapest.

—— (1982*a*). Anorectic agents and satietin, and endogenous inhibitor of food intake. In *CNS Pharmacology – Neuropeptides.* (eds. Yoshida, H., Hagihari, Y., Ebashi, S.) pp. 147–62. Pergamon Press, Oxford.

—— (1982*b*). Satietin: endogenous regulation of food intake. In *Regulatory peptides: from molecular biology to function* (eds. Costa, E., Trabucchi, M.) pp.501–9. Raven Press, New York.

—— (1982*c*). Satietin: a centrally acting potent anorectic substance with a long-lasting effect in human and mammalian blood. *Pol. J. Pharmac. Pharm.* **34**, 3–16.

—— and Knoll, B. (1958). Mehtode zur Untersuchung der spezifisch depressiven Wirkung von 'Tranquilizern' auf das Zentralnervensystem. *Arznemittel-Forsch.* **8**, 330–3.

—— —— (1959). Methode zur Untersuchung der spezifisch depressiven Wirkung von 'Tranquilizern' auf das Zentralnervensystem. 2. Mitteilung. *Arzneimittel-Forsch.* **9**, 633–6.

—— —— (1964). The cumulative nature of the reserpine effect and the possibilities of inhibiting cumulation pharmacologically. *Arch. Int. Pharmacodyn.* **148**, 200–16.

Levine, A., Morley, J. E. (1981). Reduction of feeding in rats by calcitonin. *Brain Res.* **222**, 187–91.

Lotter, E. C., Krinsky, R., McKay, J. M., Trenner, C. M., Porte, D. Jr., and Woods, C. S. (1981). Somatostatin decreases food intake of rats and baboons. *J. Comp. Physiol. Psychol.* **95**, 278–87.

Malaisse-Lagae, F., Carpentier, J. L., Patel. Y. C., Malaisee, W. J., and Orci, L. (1977). Pancreatic polypeptide: a possible role in the regulation of food intake in the mouse. Hypothesis. *Experientia* **33**, 915–18.

Morgane, P. J. (1961). Alterations in the feeding and drinking behaviour of rats with lesions in globi pallidi. *Am. J. Psychol.* **201**, 420–28.

Morley, J. E. and Levine, A. S. (1982). Corticotropin releasing factor, grooming and ingestive behaviour, *Life Sci.* **31**, 1459–60.

Nagy, J., and Kalasz, H., and Knoll, J. (1982). An improved method for the preparation of highly purified satietin samples from human serum. *Pol. J. Pharmac. Pharm.* **34**, 47–52.

—— —— —— (1983). Isolation and characterization of a highly selective anorexogenic substance by chromatography and electrophoresis. In *Chromatography and mass spectrometry in biomedical sciences* (ed. A. Frigerio) Vol. 2, pp. 421–32. Elsevier, Amsterdam.

Novin, D., Wyrwiczka, W., and Bray, G. A. (eds) (1976). *Hunger: Basic mechanisms and clinical implications.* Raven Press, New York.

Oomura, Y., Shimizu, N., Inokuchi, A., Sakata, T., Arase, K., Fujimoto, M., Fukushima, M., and Tsutsui, K. (1982). Regulation of feeding by blood-borne hunger and satiety substances through glucose-sensitive neurones. In *12th Annual Meeting of Society for Neuroscience*, Minnesota, USA, Vol. 8, Part 1, p. 10.

Reichelt, K. L., Foss, I., Trygsted, O., Edmison, P. D., Johansen, J. H., and Boler, J. B. (1978). Humoral control of appetite. II. Purification and characterization of an anorexogenic peptide from human urine. *Neuroscience* **3**, 1207–11.

Sandor, G. and Knoll, J. (1982). The selectivity of the anorexogenic effect of satietin. II. The ineffectiveness of satietin on the water intake of food deprived rats. *Pol. J. Pharmac. Pharm.* **34**, 25–32.

Schally, A. V., Redding, R. W., Lucien, H. W. and Meyer, J. (1967). Enterogastrone inhibits eating by fasted rats. *Science* **157**, 210–11.

Schulman, J. L., Carleton, J. L., Whitney, G. and Whitehorn, J. C. (1957). Effect of glucagon on food intake and body weight in man. *J. Appl. Physiol.* **11**, 419–21.

Silverstone, T. (ed.) (1982). *Drugs and appetite.* Academic Press, London.

Stellar, E. (1954). The psychology of motivation. *Psychol. Rev.* **61**, 5–22.

Timar, J. and Knoll, J. (1982). The selectivity of the anorectic effect of satietin. III. The ineffectiveness of satietin on metabolic rate, body temperature and blood pressure. *Pol. J. Pharmac. Pharm.* **34**, 33–9.

Ungerstedt, U. (1971). Adipsia and aphagia after 6-hydroxydopamine induced degeneration of the nigrostriatal dopamine system. *Acta physiol. scand.* (Suppl. 367) **82**, 95–122.

Vijayan, E. and McCann, S. M. (1977). Suppression of feeding and drinking activity in rats following intraventricular injection of thyrotropin releasing hormone. *Endocrinology* **100**, 1727–30.

Wayner, M. J. and Oomura, Y. (eds.) (1975). Central neural control of eating and obesity. *Pharmac. Biochem. Behav.* **3**, Suppl. 1.

Woods, S. C., West, D. B., Stein, L. J., McKay, L. D., Lotter, E. C., Porte, S. G., Kenney, N. J., and Porte, D. Jr. (1981) Peptides and the control of meal size. *Diabetologia* **20**, 305–13.

Yen, T. T., Dalló, J., and Knoll, J. (1982). The selectivity of the anorectic effect of satietin. IV. The ineffectiveness of satietin on the sexual performance of male CFY rats. *Pol. J. Pharmac. Pharm.* **34**, 41–5.

6

Clinical use of anorectic drugs in obesity

J. F. MUNRO AND C. ABEL

INTRODUCTION

This article assess the clinical use of those antiobesity drugs currently available. Any evaluation of their therapeutic significance requires consideration, not only of their efficacy but also of their side-effects (Douglas and Munro 1982), though these in turn must be compared with the risks of obesity itself and the hazards of the alternative forms of therapy (Garrow 1981).

OVERALL EFFICACY

In 1971, a working party, set up by the Federal Drug administration, reported their analysis of over 170 studies involving more than 1000 subjects treated either with an antiobesity drug or with a placebo in the double-blind mode (Scoville 1973). The working party found that the mean additional weight loss that could be attributed to antiobesity drugs was greeted by some as suggesting that drugs had a major role to play in the management of the obese; certainly, a sustained weight loss of 200 g per week would result in an annual weight reduction of over 10 kg. Others, however, emphasized that many of the studies that the working party had assessed were of short duration, often only four weeks, and that weight reduction would not be sustained. They implied that the additional benefit for drug therapy could not be justified.

Certainly, the fact that weight loss plateaus rapidly is borne out by the studies that have been undertaken in Edinburgh of subjects with refractory obesity. We consider that subjects have developed refractory obesity if they have been attending an obesity clinic for a period of not less than 12 months and have failed to lose weight during the preceding three months. In order to produce a 'clinical model' which would give predictable results, female subjects only are studied. The administration of placebo/drug to such patients, will produce zero weight change (Fig. 6.1). The mean weight loss achieved by a variety of antiobesity drugs is frankly disappointing, ranging from 1.2 kg to 4.2 kg during a 12 week period. Moreover, this mean weight loss is usually achieved within eight weeks of initiating therapy, weight change thereafter plateauing (Munro, Seaton, and Duncan 1966). The solitary exception is fenfluramine where there is the suggestion that weight loss is sustained throughout the 12-week period (Munro 1973).

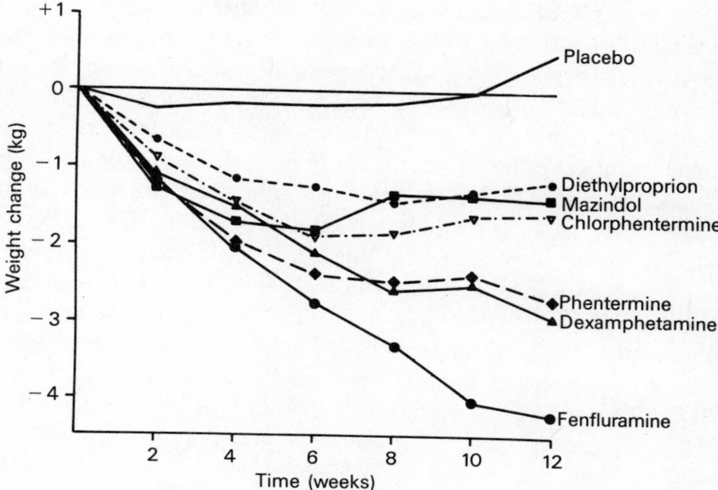

FIG. 6.1 Mean weight losses achieved with various anorectic drugs in female patients with refractory obesity (Munro 1973). (Adapted from and reproduced by permission of the Symposium Editor, Royal College of Physicians, Edinburgh).

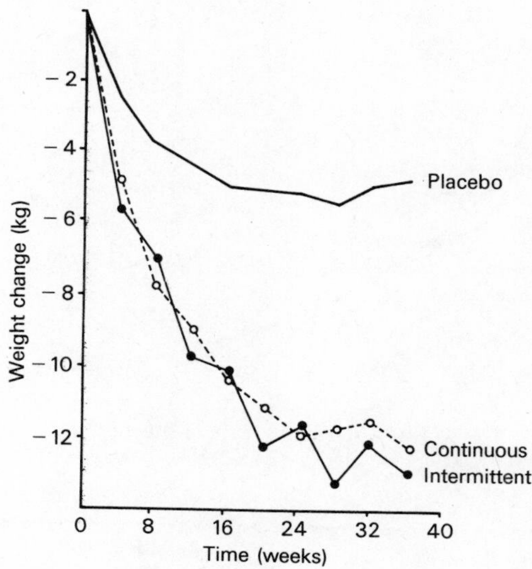

FIG. 6.2 Mean weight losses achieved with phentermine 30 mg given continuously and intermittently and compared with placebo. n = 64 (number of subjects who completed study). Mean age = 38 years. Percentage of subjects in excess of ideal weight = 54 per cent. (From Munro *et al.* (1968). Reproduced by permission of the *British Medical Journal*).

Although a useful clinical model, better results can be achieved if an anti-obesity drug is exhibited before refractory obesity has developed. Given at time of initial referral to a hospital out-patient obesity clinic, mean weight loss may be achieved for a period of up to 36 weeks (Munro, MacCuish, Wilson, and Duncan 1968) (Fig. 6.2). In this particular study, phentermine given continuously, and also administered intermittently, was significantly more effective than a matching placebo. In a further study (Fig. 6.3), intermittent phentermine was shown to be as effective as continuous fenfluramine whereas intermittent fenfluramine produced a significantly less effective response (Steel, Munro, and Duncan 1973). It seems possible that this can be attributed to the withdrawal depression that occurs when fenfluramine is stopped abruptly (Steel and Briggs 1972). For this reason, it is recommended that the drug should be given for a reasonably lengthy period of time, the dose being built up and reduced stepwise. On the other hand, phentermine and other similar drugs such as diethylpropion and mazindol, appear to be of comparable efficacy when given intermittently as distinct from continuously, and intermittent therapy is to be preferred on the grounds that not only is it cheaper, but also presumably, is less likely to result in drug dependence. One regime is to give the active preparation for four weeks every second month, though possibly alternative regimes may be as, or even more, effective.

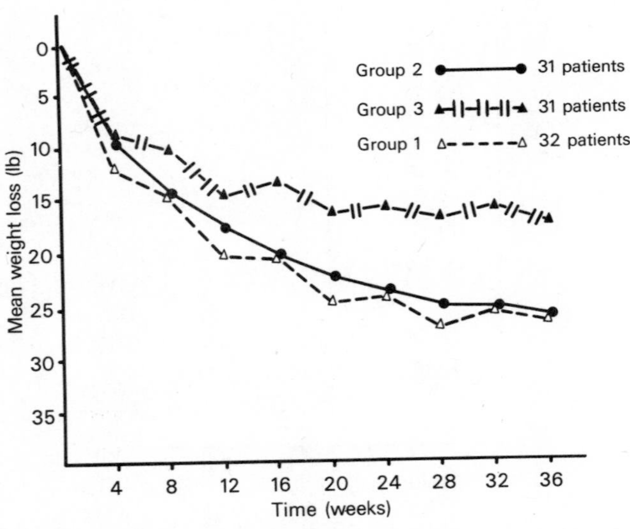

FIG. 6.3 Mean weight losses achieved with phentermine given intermittently compared with fenfluramine given continuously and intermittently. Group 1: Intermittent treatment with phentermine; Group 2: Continuous treatment with fenfluramine; Group 3: Intermittent treatment with fenfluramine. (From Steel *et al.* 1973. Reproduced by permission of the Symposium Editor, Royal College of Physicians, Edinburgh).

VARIABILITY IN SUBJECT RESPONSE

Consideration of mean weight losses conceals the fact that individual responses to antiobesity drugs vary enormously, even in patients with refractory obesity (Table 6.1). With each of the active drugs some subjects achieved substantial weight loss, and it follows that others failed to achieve any weight loss at all. Indeed, it might be argued that it is more important to 'choose' the right subject than it is to select 'the best drug'. Indeed, the ultimate choice of drug depends not so much upon efficacy as upon other factors such as relative cost, and the incidence and nature of adverse effects.

Unfortunately, there is no way of selecting those subjects who are likely to do well. Patient compliance must be a factor, but is not the only one. Among the others are individual differences in drug absorption and metabolism. In order to test this possibility, fenfluramine was administered to a group of subjects with refractory obesity, the dose being increased stepwise either until effective weight loss was achieved or until side effects developed. Once a 'plateau' dose of fenfluramine had been reached, plasma fenfluramine and norfenfluramine values were assayed on at least two occasions, and a mean concentration obtained for each subject. Subsequent analysis of data showed that there was a relationship between plasma fenfluramine values and weight loss (Innes, Watson, Ford, Munro, Stoddart, and Campbell 1977) (Fig. 6.4). Subjects who failed to achieve a mean plasma fenfluramine concentration of 100 ng/ml lost only 2 kg whereas those achieving concentrations in excess of 200 ng/ml lost almost 9 kg in the 20-week study (in other words, they lost an extra 1 lb per week). The situation, however, is not as simple as this figure might imply. Further experience has shown that some subjects fail to lose weight even when very high plasma fenfluramine concentrations have been obtained (Preston, Ford, Munro, and Campbell 1979). Indeed, a further clinical study failed to show a relationship between weight loss and plasma fenfluramine concentrations (Pietrusko, Stunkard, Brownell, and Campbell 1982). Likewise, in a study of similar design, using phentermine, we have failed to find a relationship between weight loss and plasma phentermine concentrations (Douglas, Douglas, Robertson, and Munro 1983). It follows that at the end of the day, individual response to an anti-obesity drug can only be evaluated on a basis of trial and error.

TABLE 6.1 *Mean and maximum weight losses in kg achieved in female subjects with refractory obesity undergoing 12 weeks therapy*

Drug	Dose (mg/day)	Mean	Max.
Diethylpropion	100	1.2	4.6
Mazindol	2	1.4	4.5
Phentermine	30	2.7	6.8
Dexamphetamine	15	2.9	6.8
Fenfluramine	80	4.2	9.1

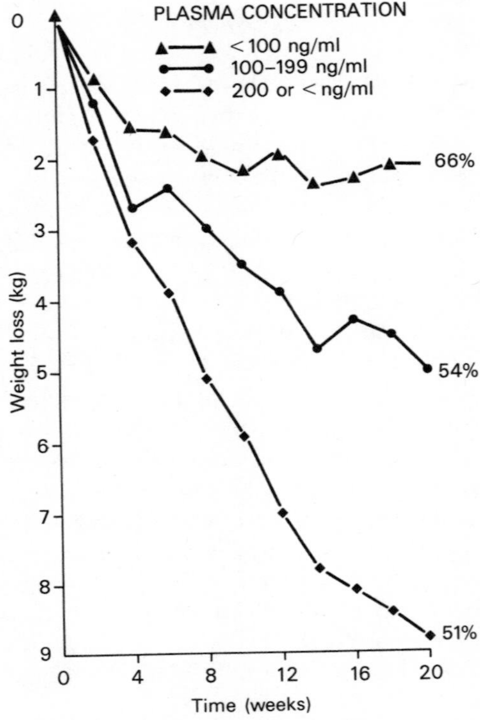

FIG. 6.4 Showing the relationship between weight loss and plasma fenfluramine concentrations in female subjects with refractory obesity. $n = 41$ (Number of subjects who completed study. Mean age = 46 years. Percentage of subjects in excess of ideal weight = 57 per cent). (From Innes *et al.* 1977. Reproduced by permission of the *British Medical Journal*).

EFFECT OF DISCONTINUATION OF DRUG THERAPY

An analysis of those cross-over studies which compare an active drug with placebo confirm the clinical impression that once the drug is discontinued, weight regain is the rule. It follows that drug therapy can be best justified if there is a short term need for weight reduction. One obvious situation is in the subject who requires an elective operation, but has failed to reach a weight which the surgeon considers acceptable. In these circumstances, the administration of the drug will often assist the subject to achieve a target weight and thereby undergo relatively risk-free surgery. Where there is no short-term justification for weight loss, the administration of a drug is hard to justify. Indeed, it might be argued that in the long term it is to the subject's detriment. One study has compared fenfluramine administered 'alone' with fenfluramine given in conjunction with a behavioural modification programme, and with behavioural modification without pharmacological treatment. The study had been undertaken to test the hypothesis that combined therapy might produce more impressive weight loss. In the event, drug treatment alone proved to be superior to behavioural modification, and comparable to that of drug therapy with behavioural modification (Stunkard, Craighead, and O'Brien 1980). However, during a 12-month follow-up period, those subjects who had not received the drug, regained the least weight so that those who have received fenfluramine either with or without behavioural modification, finished up heavier than those who had not been treated with the drug. Certainly, there is no evidence to suggest that an antiobesity drug will 'help to improve eating habits'. Indeed, the converse may be the case and this further emphasizes the fact that a short course of drug treatment should only be prescribed when there is a specific purpose in sight.

LONG-TERM THERAPY

It is customary to discontinue treatment once the weight loss has stopped, possibly on the grounds that this implies that tolerance to the drug has developed. However, there is anecdotal evidence to suggest that mazindol (Enzi, Baritussio, Machion, and Cepaldi 1976), diethylpropion (Craddock 1977; Matthews 1975) and phentermine (Smith 1962), given on a long-term basis, will not only produce weight loss but will also prevent subsequent weight regain. One impressive study with fenfluramine has shown that after a mean weight loss of 11.5 kg in six months, weight change then plateaued while the drug was continued for a further six months, but weight regain occurred rapidly with discontinuation of the drug treatment (Hudson 1977). Figure 6.5 shows the result of a further study in which 42 subjects, each of whom had lost at least 6 kg in weight during six months treatment with fenfluramine, were changed over in a double-blind mode to continue treatment either with fenfluramine pacaps or matching placebo. Of the 21 subjects who were converted to placebo, 19 rapidly regained weight. Six subjects receiving fenfluramine also regained weight and this weight regain might suggest true drug tolerance. Seven other subjects had to be withdrawn for a variety of reasons but eight subjects failed to regain weight, indeed some continued to lose weight (Douglas, Gough, Preston, Frazer, Haslett, Chalmers, and Munro 1983). It thus seems fair to assume that at least a proportion of subjects continuing drug therapy will produce a satisfactory reduction in weight which can be maintained for the duration of drug treatment. Unfortunately, during the run-in period to the cross-over, one subject, not included in the study itself, developed pulmonary hypertension which corrected when the drug was discontinued and recurred on rechallenge (Douglas, Munro, Kitchin, Muir, and Proudfoot 1981). Although it is possible that this complication is infrequent, the inherent risks of long-term treatment either with fenfluramine or with one of the alternative antiobesity agents, has yet to be properly evaluated. Certainly, if long-term therapy can be justified at all, it is in those patients with complicated obesity suffering from conditons such as diabetes or hypertension which will themselves be alleviated by effective long-term weight loss. Indeed, one drug, metformin, is already being used at least in part because of its weight stabilizing effect in the obese insulin-independent diabetic (Clarke and Duncan 1968).

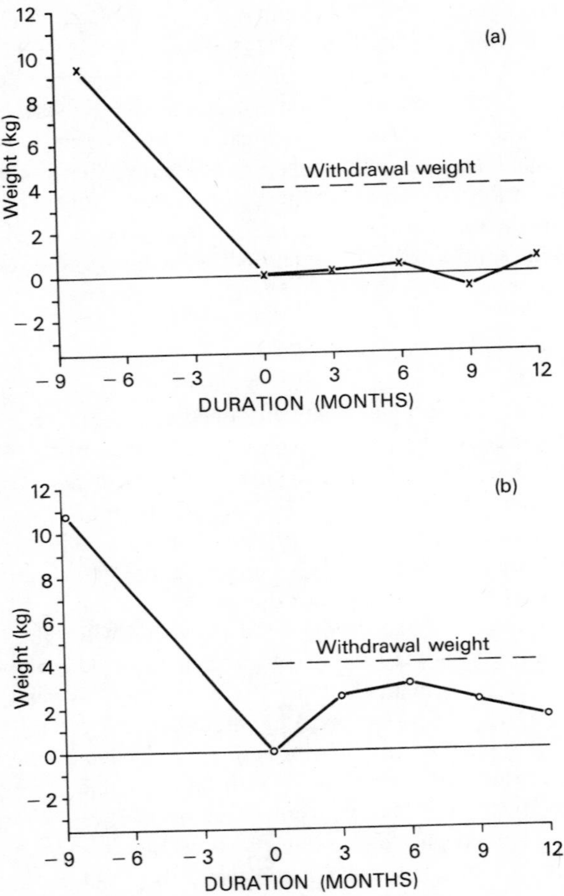

FIG. 6.5 Mean weight change in subjects initially treated with fenfluramine and then crossed over to continue with (a) fenfluramine therapy of (b) placebo.

THE FUTURE

Clearly, at present, the major drawback to the pharmacological treatment of obesity is the problem of weight regain. If this could be prevented, using some non-pharmacological approach, the indications for drug treatment might be considerably enhanced. Recently, it has been shown that weight regain that occurs following dental splinting can be minimized by applying the nylon rope round the subject's abdomen once substantial weight loss has occurred (Garrow and Gardiner 1981). It seems possible that this concept might be used to ensure that weight loss achieved by pharmacological intervention is similarly not regained.

CONCLUSIONS

(1) Currently available antiobesity drugs will produce an additional mean weight loss of approximately 200 g (½ lb) per week for a limited period of time.

(2) Individual response to drugs varies enormously, some subjects losing substantially more than the mean weight loss.

(3) Once the drug is discontinued, weight regain is the rule. It follows that drug therapy can be best justified if there is a short term need to achieve weight loss.

(4) In some subjects, weight regain may be prevented by giving the drug long-term, but the complications of this therapy have yet to be evaluated. If it can be justified at all, it is in those subjects with complicated obesity.

(5) The development of a non-pharmacological way of preventing weight regain following drug therapy would enormously enhance the potential usefulness of an antiobesity agent.

References

Clarke, B. F. and Duncan, L. J. P. (1968). Comparison of chlorpropamide and metformin treatment on weight and blood glucose response of uncontrolled obese diabetics. *Lancet* i, 123–6.

Craddock, D. (1977). The free diet – 150 cases. *Int. J. Obes.* **1**, 127–34.

Douglas, A., Douglas, J. A., Robertson, C. E., and Munro, J. F. (1983). Plasma phentermine levels, weight loss and side effects. *Int J. Obes.* **7**, 591–5.

Douglas, J. G. and Munro, J. F. (1982). Drug treatment and obesity. *Pharmac. and Therap.* **18**, 351–3.

—— Munro, J. F., Kitchin, A. H., Muir, A. L. and Proudfoot, A. T. (1981). Pulmonary hypertension and fenfluramine. *Br. Med. J.* **283**, 881–3.

—— Gough, J., Preston, P. G., Frazer, I., Haslett, C., Chalmers, S. R., and Munro, J. F. (1983). Longer term efficacy of fenfluramine in obesity. *Lancet* i, 384–6.

Enzi, G., Maritussio, A., Machion, E., and Cepaldi, G. (1976). Short-term and long-term clinical evaluation of a non-amphetamine anorexic (mazindol) in the treatment of obesity. *J. Int. Med. Res.* **4**, 305–18.

Garrow, J. S. (1981). *Treat obesity seriously.* Churchill Livingstone, Edinburgh.

—— Gardiner, G. T. (1981). Maintenance of weight loss in obese patients after jaw wiring. *Br. Med. J.* **282**, 858–60.

Hudson, K. D. (1977). The anorectic and hypotensive effect of fenfluramine in obesity. *J. R. Coll. Gen. Pract.* **16**, 6–9.

Innes, J. A., Watson, M. L., Ford, M. J., Munro, J. F. Stoddart, M. E., and Campbell, D. B. (1977). Plasma fenfluramine levels, weight loss and side effects. *Br. Med. J.* **2**, 1322–5.

Matthews, P. A. (1975). Diethylpropion in the treatment of obese patients seen in general practice. *Curr. Ther. Res.* **17**, 340–6.

Munro. J. F. (1973). The management of obesity. In *Anorexia nervosa and obesity*. Royal College of Physicians, (Edinburgh), Symposia No. 42, pp. 100–9.

—— Seaton, D. A. and Duncan, L. J. P. (1966). Treatment of refractory obesity with fenfluramine. *Br. Med. J.* **2**, 624–5.

—— McCuish, A. C., Wilson, E. M., and Duncan, L. J. P. (1968). Comparison of continuous and intermittent anorectic therapy in obesity. *Br. Med. J.* **1**, 352–4.

Pietrusku, R., Stunkard, A., Brownell, K., and Campbell, D. B. (1982). Plasma fenfluramine levels, weight loss and side effects: a failure to find a relationship. *Int. J. Obes.* **6**, 567–71.

Preston, P. G., Ford, M. J., Munro, J. F., and Campbell, D. B. (1979). The variable response to fenfluramine in obesity. *Int. J. Obes.* **3**, 359–61.

Scoville, B. A. (1973). Review of amphetamine-like drugs by the Food and Drug Administration: Clinical data and value judgement. In *Obesity in perspective*, pp. 441–3. Proceedings of the Fogarty Conference, US Government, Washington.

Smith, R. C. F. (1962). The long term control of obesity using sustained release appetite suppression. *Br. J. Clin. Pract.* **16**, 6–9.

Steel, J. M. and Briggs, W. (1972). Withdrawal depression in obese patients after fenfluramine treatment. *Br. Med. J.* **3**, 284–5.

—— Munro, J. F., and Duncan, L. J. P. (1973). A comparative trial of different regimes of fenfluramine and phentermine in obesity. *Practitioner* **211**, 232–6.

Stunkard, A. J., Craighead, L. W., and O'Brien, R. (1980). Controlled trial of behaviour therapy, pharmacotherapy and their combination in the treatment of obesity. *Lancet* **ii**, 1045–7.

7

Psychotropic drugs, appetite, and body weight

TREVOR SILVERSTONE

INTRODUCTION

Psychotropic drugs are substances that, by virtue of their action on the brain, affect normal or abnormal psychological processes. Among such processes are those governing the perceptions of hunger and appetite, which in turn influence the regulation of food intake and body weight. It is my purpose in this review to describe the ways in which drugs, administered primarily to alter abnormal psychological processes (as in the treatment of psychopathological states), can at the same time affect appetite, and thereby lead to alterations in body weight. Changes in weight brought about by psychotropic drugs can adversely influence compliance, consequently increasing the risk of relapse. Furthermore, drug-induced obesity, in common with other forms of overweight, is accompanied by an increase in morbidity and mortality (Royal College of Physicians 1983).

This relationship between psychotropic drugs, and appetite and body weight is also of potential interest to psychiatrists, because an understanding of the ways these drugs act to produce such effects can throw light on the neurochemical mechanisms underlying those disease states for which the drugs are being administered (Silverstone 1983).

Before proceeding further, I would like to clarify what I mean by the terms 'appetite' and 'hunger'. Appetite is the desire for a given food at a particular time; it is influenced by, but is not synonymous with, hunger. Such desire may occur at the sight or smell of a delectable food. Hunger, on the other hand, is the psychological expression of those physiological changes that accompany prolonged food deprivation (for a fuller discussion of the differentiation between hunger and appetite, see Silverstone 1982).

While the effects of psychotropic drugs on food intake are likely to be largely a result of their effects on those central physiological processes which are concerned with hunger (Morley and Levine 1983), they can also influence food intake through affecting psychomotor function or altering underlying mood. The psychotropic drugs I shall be considering will include psychostimulants, antidepressants, antipsychotics and lithium.

PSYCHOSTIMULANTS

The best-studied member of this class of drugs is amphetamine. Amphetamine was originally synthetized to provide a cheaper synthetic substitute for ephedrine (Alles 1927). Within a short time, its stimulant and euphoriant properties were recognized, and it was recommended for use in the treatment of narcolepsy and for alleviating depressive symptoms (Printzmetal and Bloomberg 1935). The finding that certain patients being treated with amphetamine for narcolepsy lost weight led to the suggestion that it might be of value in the treatment of obesity (Davidoff and Reifenstein 1937). Although the use of amphetamine in clinical practice is now strictly limited, largely because of its abuse potential, its particular constellation of actions makes it of considerable interest to study from the clinical psychopharmacological point of view. It has been shown in rats that amphetamine releases two neurotransmitters, dopamine (DA) and noradrenaline (NA), from presynaptic neurones (Carlsson 1970). It has further been demonstrated that the stimulant activity of low doses of amphetamine in rats is dependent on DA pathways whereas the anorectic action is not (Burridge and Blundell 1979).

What is the situation in humans? One way of examining this question is to measure both the stimulant and anorectic activity of the dextro-isomer of amphetamine (*d*-AMP) in the presence of relatively specific receptor-blocking compounds. We have done this using pimozide (PMZ) as a DA receptor blocker and thymoxamine (TMX) as an NA receptor blocker. PMZ had no effect on *d*-AMP-induced anorexia whereas it attenuated *d*-AMP arousal (Silverstone, Fincham, Wells, and Kyriakedes 1980). TMX on the other hand attenuated *d*-AMP anorexia, but not *d*-AMP arousal. This suggests that *d*-AMP arousal in humans, as in rats, is mediated through central DA pathways whereas *d*-AMP anorexia may be mediated through NA pathways.

Amphetamine has been shown to improve mood, albeit for only a relatively short time, in a proportion of depressed patients (Kiloh, Neilson, and Andrews 1974; Checkley 1978). It is well recognized that anorexia is a prominent feature of severe depressive states (Lewis 1934), and that clinical improvement is frequently heralded by an improvement in appetite (Russell 1960). The question then arises; during *d*-AMP induced improvement of mood in a depressed patient, is appetite improved, as might occur as an accompaniment of the alleviation of depression, or does it remain diminished, due to the anorectic properties of amphetamine? To answer these questions we administered 15 mg methylamphetamine (*m*-AMP) or sterile water intravenously under double-blind conditions to 21 seriously depressed patients (T. Silverstone and J. C. Cookson, in preparation). Seven patients responded to *m*-AMP (but not to placebo) with a marked elevation of mood, as assessed by appropriate visual anologue scales. Of these seven, six experienced a concommitant increase in subjective hunger. This finding indicates that, in some patients with depressive illness, DA pathways appear to be intact (as reflected by the elevation of mood occuring in response to *m*-AMP), whereas the activity of NA pathways is diminished, as evidenced by the attenuation of the expected anorectic response to *m*-AMP. This view is consistent with other observations relating to a blunting of central NA responsiveness in depressive illness (Checkley and Crammer 1977).

ANTIDEPRESSANTS

As we have already seen improvement in mood in depressed patients is accompanied by an increase in appetite. Thus, it would be expected that effective treatment with an antidepressant drug might well be associated with a gain in weight. However, any gain in weight over and above that associated with general clinical improvement could lead to undesirable consequences. A potential for excess weight gain has been particularly remarked upon with the antidepressant drug amitriptyline. Arenillas (1964) was the first to draw attention to the frequency with which patients on amitriptyline gained weight, and he pointed out that this weight gain appeared to be the result of a craving for carbohydrate-containing foods. The suggestion that this carbohydrate-carving might be due to an effect of amitriptyline on insulin was postulated by Winston and McCann (1972); they reported two patients on a combination of antidepressants who had gained weight and in whom circulating insulin levels appeared to be elevated.

Paykel, Muller, and de la Vergne (1973) took the matter further. They examined the relationship of amitriptyline to weight gain which was over and above that which might have been expected simply as a result of clinical improvement. They found that patients who were being maintained on prophylactic amitriptyline, having previously recovered from a depressive illness, gained significantly more weight than a matching control group who were being maintained on placebo. As before, the increased weight appeared to be secondary to an increase in a desire for carbohydrate-containing foods.

However, there was no correlation between any change in either plasma-glucose or insulin tolerance and carbohydrate-craving. Similarly, Nakra, Rutland, Verma, and Gaind (1977) observed no alteration in glucose tolerance, or in fasting or peak insulin levels in normal volunteer subjects given amitriptyline 50 mg twice daily for 28 days.

One possible mechanism by which amitriptyline might cause an increase in weight is through blockade of central serotonergic receptors, a propety which the drug is known to possess. (Maj, Lewandowska, and Rawlow 1979; Peroutka and Snyder 1980). Certainly, other central serotonergic receptor-blocking compounds such as cyproheptadine cause marked increase in appetite and body weight (Silverstone and Schuyler 1975). Furthermore, fenfluramine which releases serotonin from presynaptic neurones (Garattini and Samanin 1976) is a potent anorectic compound in humans (Kyriakides and Silverstone 1979). Similarly, tryptophan, a precursor of serotonin sometimes used as a treatment for depression, has also been found to reduce food intake in normal human subjects (G. Smith and T. Silverstone, in preparation).

ANTIPSYCHOTIC DRUGS

Phenothiazine compounds, chlorpromazine in particular, have long been noted to cause pronounced weight gain. For example, Amidsen (1964), in a retrospective study, found that some 80 per cent of 179 patients who had been treated with chlorpromazine gained weight; this weight gain was considered to be excessive in over a quarter of the patients, who exceeded their ideal body weight by 25 per cent or more. More recently, Harris and Eth (1981) conducted a prospective survey on an acute admission ward. Eleven of the 13 patients who received chlorpromazine gained weight, as did 9 of the 11 treated with thiothixine. It is likely that such weight gains are the direct result of a drug-induced increase in appetite, as Robinson, McHugh and Follstein (1975) observed a clear-cut dose relationship; those patients who were receiving a higher dose of chlorpromazine had higher hunger ratings and a greater weight gain than those receiving a lower dose.

It is when it comes to long-term medication in the management of chronic schizophrenia that the question of drug-induce obesity assumes major clinical importance. For maintenance of their psychological and social well-being, it is essential that such patients continue to receive antipsychotic medication on a regular basis; for many this is administered as a depot injection. Any increase in the prevalence of obesity brought about by the medication is likely to reduce compliance, particularly in younger women, because in our society, as Hilda Bruch so pithily put it, 'Slenderness is next to Godliness' (Bruch 1974). In addition obesity, as has already been stated, is associated with an increased risk of morbidity from such conditions as diabetes, cardio-vascular disease, and locomotor disorders.

Hence there are cogent medical as well as social reasons for avoiding drug-related obesity. How common is obesity among patients receiving regular depot antipsychotic drugs? Johnson and Breen (1979) weighed regularly 132 patients receiving either fluphenazine decanoate (Modecate) or flupenthixol decanoate (Depixol) over a six-month period. 28 per cent of the patients gained over 3 kg during this time. There was no difference between fluphenazine and flupenthixol in this respect. Marriot, Pansa, and Hiep (1981) conducted a prospective study of 110 patients starting on fluphenazine decanoate and who remained on it for a at least 6 months. They found a rather lower incidence of obesity; less than 15 per cent of patients gained even 3 lb. On the basis of their findings they concluded that obesity was not likely to prove much of a problem among patients receiving depot fluphenazine maintenance treatment.

In order to re-examine this question of whether or not long-term depot injections of fluphenazine and/or flupenthixol are associated with a worryingly high frequency of obesity, we weighed and measured 228 patients (106 female, 122 male) who were attending for regular depot injections (Smith and Silverstone, in preparation). The degree of any obesity present was graded according to a scheme suggested by Garrow (1981) based on the ratio of weight to height ($Wt(kg)/Ht(m)^2$). A ratio of 20–25 is classified on Grade 0 (desirable), 25–30 Grade 1 (mild), 30–40 Grade 2 (clinically relevant), >40 Grade 3 (crippling). 76 (33 per cent) of our patients were in the 'clinically relevant obesity' category, with 11 of them being in the 'crippling' category. For comparison, only 5 per cent of women and 4 per cent of men who are at work exhibit Grade 2 obesity and none exhibit Grade 3 (Garrow 1981); and in the population at large, no more than 6 per cent of men and 8 per cent of women show Grade 2 obesity (Office of Population Censuses and Surveys 1981). Thus, we are faced with the disturbing conclusion that long-term maintenance treatment with depot injections of fluphenazine and flupenthixol is associated with at least a four-fold increase in the prevalence of clinically relevant obesity. It is unlikely that such a sharp increase in the prevalence of obesity is a necessary accompaniment of successful treatment. A recent double-blind comparative trial of haloperidol decanoate and fluphenazine decanoate revealed that weight gain was significantly less likely to occur with haloperidol decanoate although both preparations were equally effective in alleviating psychotic symptoms (Wistedt and Person 1982). This finding is consistent with our own observation that dopaminergic blockade has no influence on subjective hunger (Silverstone *et al.* 1980). Chlorpromazine, fluphenazine and flupenthixol all have other neuropharmacological actions apart from dopamine receptor blockade; these include a serotonergic receptor-blocking activity which could well underlie their appetite and weight-increasing potential.

LITHIUM

Weight gain is a commonly reported side-effect among patients recieving long-term lithium prophylaxis for manic-depressive illness. Here, too, weight gain is likely to influence compliance adversely with a consequent increase in the likelihood of relapse. The reported frequency of such weight gain varies. Schou, Baastrup, Grof, Weis, and Angst (1970) reported a weight gain of at least 5 kg in 11 per cent of their patients. Dempsey, Dunner, Fieve, Farkas, and Wong (1976) stated that approximately one third of their patients gained some weight on lithium, with 'several' gaining in excess of 5 kg. Vendsborg, Bech, and Rafaelsen (1976) found a mean weight gain of 7.5 kg in 64 per cent of their patients. Self-reports of weight gain range from 20 per cent in one large series (Vestergaard, Amdisen, and Schou 1980) to 36 per cent in a much smaller series (Duncavage, Nasr, and Altman 1983).

We have recently determined the prevalence of obesity among 59 regular attenders at our lithium clinic. All the patients had been on lithium for at least three months, many for years, and all had plasma lithium levels within the recommended range. Of these 59 patient eleven (19 per cent) exhibited Grade 2 obesity, with one being Grade 3. Thus, the prevalence of obesity among regular lithium takers is two to three times higher than in the general population.

The mechanism by which lithium induces weight gain is poorly understood. Vendsborg *et al.* (1976) examined the association between increased appetite, increased thirst, and weight gain among 70 regular attenders at a lithium clinic. Approximately one third of the patients considered that their appetite had increased while on lithium and of those, the great majority showed an increase in weight. However, the mean weight gain among those who did not report any increase in appetite, while less, was not significantly so. The great majority of patients (80 per cent) remarked on an increases thirst. Here there was a significant correlation between the subjective experience of thirst and weight gain. The mean weight gain among those experiencing little increase in thirst was 1.9 kg compared to 7.0 kg among those showing a moderate increase. Furthermore, the 21 patients whose polyuria and polydipsia were severe enough for them to be categorized as suffering from frank diabetes insipidus, gained the most weight (8.3 kg on average). It would therfore appear that at least a proportion of the weight gain observed in patients on lithium is secondary to an increased consumption of high calorie drinks, a consequence of their marked thirst. An obvious simple preventive measure would be to advise patients on lithium to assuage their thirst with plain water or low-calorie beverages, rather than with drinks containing a large number of calories.

CONCLUSIONS

In this review I have presented some examples of the ways in which psychotropic durgs, prescribed to ameliorate a wide range of psychiatric conditions, can influence appetite and body weight. And how thereby this can significantly affect the outcome of such treatment by lessening compliance and producing secondary morbidity. Furthemore, I have illustrated how detailed pharmacological examination of these effects on appetite and food intake, when undertaken with a simultaneous examination of their effects on mental state, can throw light on the pathogenesis of certain psychiatric conditions.

Thus, I would argue that further studies of the interrelationships between appetite, psychopathology, and psychopharmacology is highly likely to prove fruitful in increasing our understanding of psychiatric illness.

References

Alles, G. A. (1927). Comparative physiological action of phenylethanolamine. *J. Pharmac.* **32**, 121–6.

Amdisen, A. (1964). Drug-produced obesity: experiences with chlorpromazine, perphenazine and clopenthixol. *Dan. Med. Bull.* **11**, 182–9.

Arenillas, L. (1964). Amitriptyline and body weight. *Lancet* **i**, 432–3.

Bruch, H. (1974). *Eating disorders.* Routledge and Kegan Paul, London.

Burridge, S. L. and Blundell, J. E. (1979). Amphetamine anorexia: antagonism by typical but not atpical neuroleptics. *Neuropharmacology* **18**, 453–7.

Carlsson, A. (1970). Amphetamine and brain catecholamines. In *Amphetamine and related compounds* (eds. E. Costa and S. Garattini) pp. 289–300. Raven Press, New York.

Checkley, S. A. (1978). A new distinction between the euphoric and antidepressant effect of methylamphetamine. *Br. J. Psychiat.* **133**, 416–23.

—— and Crammer, J. L. (1977). Hormone responses to methylamphetamine in depression: a new approach to the noradrenaline depletion hypothesis. *Br. J. Psychiat.* **131**, 582–6.

Davidoff, E. and Reifenstein, E. C. (1937). The stimulating action of benzedrine sulphate. *J. Am. Med. Assoc.* **108**, 1770–4.

Dempsey, G. M., Dunner, D. L., Fieve, R., Farkas, R., and Wong, J. (1976). Treatment of excessive weight gain in patients taking lithium. *Am. J. Psychiat.* **133**, 1082–4.

Duncavage, M. B., Nasr, S. J., and Altman, E. G. (1983). Subjective side-effects of lithium carbonate: a longitudinal study. *J. Clin. Psychopharmac.* **3**, 100–2.

Garattini, S. and Samanin, R. (1976). Anorectic drugs and brain neurotransmitters. In *Appetite and food intake* (ed. T. Silverstone) pp. 83–108. Dahlem Konferenzen, Berlin.

Garrow, J. S. (1981). *Treat obesity seriously.* Churchill Livingstone, Edinburgh.

Harris, E. and Eth, S. (1981). Weight gain during neuroleptic treatment. *Int. J. Nursing Stud.* **18**, 171–5.

Johnson, D. A. W., and Breen M. (1979). Weight changes with depot neuroleptic maintenance therapy. *Acta psychiat. scand.* **59**, 525–8.

Kiloh, L. G., Neilson, M., and Andrews, G. (1974). Response of depressed patients to methylamphetamine. *Br. J. Psychiat.* **125**, 496–9.

Kyriakides, M. and Silverstone, T. (1979). A double-blind comparison of fenfluramine and dextroamphetamine on feeding behaviour in man. *Curr. Med. Res. Opinion* **6**, Suppl. 1, 180–7.

Lewis, A. (1934). Melancholia: a clinical survey of depressive states. *J. Men. Sci.* **80**, 277–378.

Maj, J., Lewandowska, A., and Rawlow, A. (1979). Central antiserotonin action of amitriptyline. *Pharmacopsychiatry* **12**, 281–5.

Marriott, P., Pansa, M., and Hiep, A. (1981). *Comp. Psychiat.* **22**, 320–5.

Morley, J. E. and Levine, A. S. (1983). The central control of appetite. *Lancet* **i**, 398–401.

Nakra, B. R. S., Rutland, R. P., Verman, S., and Gaind, R. (1977). Amitriptyline and weight gain: a biochemical and endocrinological study. *Curr. Med. Res. Opinion* **4**, 602–8.

Office of Population Censuses and Surveys (1981). OPCS monitor, ref SS 81/1.

Paykel. E. S., Muller, P. S., and De La Vergne. O. O. (1973). Amitriptyline, weight gain and carbohydrate craving: a side effect. *Br. J. Psychiat.* **123**, 501–7.

Peroutka, S. J. and Snyder, S. H. (1980). Chronic antidepressant treatment lowers spiroperidol-labelled serotonin receptor binding. *Science*, **210**, 88–90.

Printzmetal, O. O. and Bloomberg, W. (1935). The use of benzedrine for the treatment of narcolepsy. *J. Am. Med. Ass.* **105**, 2051–3.

Robinson, R., McHugh, P., and Follstein, M. (1975). Measurement of appetite disturbance in psychiatric disorders. *J. Psychiat. Res.* **12**, 59–68.

Royal College of Physicians Working Party on Obesity (1983). *J. R. Coll. Physicians* **17**, 6–65.

Russell, G. (1960). Body weight and balance of water, sodium and potassium in depressed patients given ECT. *Clin. Sci.* **19**, 327–36.

Schou. M., Baastrup, P. C., Grof, P., Weis, P., and Angst, J. (1970). Pharmacological and clinical problems of lithium prophylaxis. *Br. J. Psychiat.* **116**, 615–9.

Silverstone, T. (1982). *Drugs and appetite.* Academic Press, London.

—— (1983). The clinical pharmacology of appetite – its relevance to psychiatry. *Psychol. Med.* **13**, 251–3.

—— and Schuyler, D. (1975). The effect of cyproheptadine on hunger, calorie intake and body weight in man. *Psychopharmacologia* **40**, 335–40.

—— Fincham, J., Wells, B., and Kyriakides, M. (1980). The effect of pimozide on amphetamine-induced arousal, euphoria and anorexia in man. *Neuropharmocology* **19**, 1235–7.

Vendsborg, P. B., Bech, P., and Rafaelsen, O. J. (1976). Lithium treatment and weight gain. *Act. psychiat. scand.* **53**, 139–47.

Vestergaard, P., Amdisen, A., and Schou, M. (1980). Clinically significant side-effects of lithium treatment. A survey of 237 patients in long-term treatment. *Acta psychiat. scand.* **62**, 193–200.

Winston, F. and McCann, M. L. (1972). Antidepressant drugs and excessive weight gain. *Br. J. Psychiat.* **120**, 693–7.

Wistedt, B., and Person, T. (1982). Haloperidol decanoate and fluphenazine decanoate in schizophrenic patients. Paper presented at 13th CINP Congress, Jerusalem.

8

Do drugs have a place in the management of anorexia nervosa and bulimia nervosa?

G. F. M. RUSSELL

This article is divided into three sections. The first consists of a brief review of the different drugs that have been administered to patients with anorexia nervosa, mainly on theoretical grounds and with the aim of stimulating hunger and increasing food intake. The second section is concerned with a discussion of drug therapies aimed at a supposedly primary psychiatric disorder (e.g. depressive illness) which finds expression as an anorexic or bulimic syndrome. The third section summarizes the current empirical treatment of anorexia nervosa and bulimia nervosa. It ends with an account of two experimental studies on the suppression of symptoms of bulimia nervosa with methylamphetamine and fenfluramine.

THEORETICAL MODELS FOR ANOREXIA NERVOSA

Early ideas on the central nervous regulation of food intake were based on the results of hypothalamic lesions in animals. Bilateral lesions in the extreme lateral regions of the hypothalamus led to self-starvation and death in rats (Anand and Brobeck 1951). It was proposed that a 'feeding centre' existed in the lateral hypothalamus, whose action was opposed by a 'satiety centre' in the medial hypothalamus (Anand 1961). Later work proposed that lateral hypothalamic damage was associated with transection of the principal monoamine pathways ascending to forebrain structures by way of the medial forebrain bundle. The monoamine pathways comprised those of noradrenaline, dopamine and 5-hydroxy-tryptamine (Ungerstedt 1971*a*; Coscina 1977). It was proposed that the nigro-striatal dopamine system was the one whose destruction was most relevant to experimentally produced aphagia and adipsia in animals (Ungerstedt 1971*b*).

There have since appeared several reviews suggesting how the illness in humans, anorexia nervosa, might be due to a disturbance of neurotransmission concerned with the regulation of food intake. Many different hypotheses have been put forward, some of which are in conflict with each other, as pointed out in a critical review by Szmukler (1982). Nevertheless it is postulated that altered function of specific neurotransmitter systems gives rise to the food refusal which is characteristic of anorexia nervosa. Neurotransmitter action may be altered in the direction of overactivity or underactivity of one or more of the monoamine systems. Some authors have further proposed that the clinical disorder which characterizes anorexia nervosa may be reversed by the administration of a specific drug which would appropriately enhance or inhibit the release from axon terminals of the particular neurotransmitter, the function of which is considered altered. These postulated disturbances will be briefly described; they are summarized in Table 8.1.

Dopamine

Overactivity of the dopamine system was suggested by Barry and Klawans (1976). This idea is consistent with the observed weight gain which is induced in humans by chlorpromazine, a drug which binds tightly to dopamine receptors in the brain. Chlorpromazine enjoyed a vogue as a favoured treatment of anorexia nervosa during the 1960s (Dally and Sargant 1960, 1966; Dally 1967). On the basis of the dopamine overactivity hypothesis, Vandereycken and Pierloot (1982) administered pimozide, another selective dopamine receptor blocker. The effects of pimozide on weight gain were compared with those of placebo in patients already benefiting from a management programme of behaviour therapy. Although not conclusive, this study suggested that pimozide enhanced weight gain, but only to a modest degree. The drug had a marginal influence on the patients' abnormal attitudes measured by means of an anorexic behaviour scale. In a similar study, sulpiride was administered as a selective dopamine antagonist (Vandereycken 1984). The results were much the same as with pimozide, sulpiride tending to cause more weight gain than placebo, but the difference failed to reach statistical significance. Once again, it was not possible to establish any clear effects on the patients' attitudes measured by means of behavioural and attitudinal rating scales.

A converse hypothesis - one of progressive depletion of brain dopamine (possibly combined with noradrenaline depletion) - has been put forward by Mawson (1974). It carries less conviction, but led to a proposal to treat anorexic patients with L-dopa, a natural precursor of the catecholamine transmitters. Johanson and Knorr (1977) administered L-dopa to nine anorexic patients, and claimed that five of them benefited from the drug. But the weight gains were modest, and the study was uncontrolled so that it should be viewed as inconclusive. Acting also on the basis of a dopamine depletion hypothesis for anorexia nervosa, Harrower, Yap, Nairn, Walton, Strong, and Craig (1977) administered bromocriptine, a dopamine agonist. There resulted only slight and variable changes in weight in an open trial on eight anorexic patients, so that once again this result must be seen as disappointing.

5-hydroxytryptamine

The effect of cyproheptadine in stimulating appetite and causing weight gain has been attributed to its 5-HT antagonistic activity (Silverstone and Turner 1974), although it has other pharmacological actions. The use of this drug in anorexia nervosa has been tested in two controlled trials (Vigersky and Loriaux 1977; Goldberg, Halmi, Eckert, Casper, and Davis 1979). In neither study was cyproheptadine clearly superior to placebo.

TABLE 8.1 *Theoretical models for anorexia nervosa based on neurotransmitters*

Hypothesis for A N transmitter	neurotransmission (+,-)	Predicted	Treatment	Nature of trial
Dopamine	+	DA blocker	Pimozide	Controlled
			Sulpiride	Controlled
	–	DA precursor	L-dopa	Open
	–	DA agonist	Bromocriptine	Open
5-OH tryptamine	+?	Antagonist	Cyproheptadine	Controlled
Noradrenaline	+	α-Blocker	Phenoxybenzamine	Pilot
	–	α-Noradrenergic agonist	Clonidine	Suggestion
Endogenous opioids	+	Opioid antagonist	Naloxone	Suggestion
Oestrogen*	+	? Antagonist	Progesterone	Suggestion

The left columns indicate the various hypotheses proposed for anorexia nervosa in terms of altered neurotransmitter function.

The right columns show how drugs have been selected on the basis of these theoretical models, and the nature of any therapeutic trials undertaken (controlled, open, pilot). In the case of some drugs their use has only been suggested so far, or they have been administered for some other reason.

* Oestrogen is also tabulated for convenience even though it is not a neurotransmitter.

Noradrenaline

An increased activity of postsynaptic noradrenergic receptors was proposed by Redmond *et al.* (1976, 1977) as the cause of anorexia nervosa. This idea was derived from their experiments on monkeys in whom bilateral lesions of the locus coeruleus led to hyperphagia and weight gain. These lesions led to reduced function in a major noradrenaline system in the brain, thus interfering with a postulated adrenergic 'satiety mechanism'. These authors suggested giving alpha- or beta-adrenergic antagonists to patients with anorexia nervosa; they administered the alpha-receptor blocking drug phenoxybenzamine in one patient who gained weight apparently as a result.

In contrast, it has more recently been proposed that the anorexia of anorexia nervosa is due in part to a decrease in hypothalamic noradrenergic activity, whereas bulimic episodes in anorexia nervosa might be due to an increase of this activity. This view is based on the results of injecting noradrenaline in physiological doses through a cannula into discrete areas of the hypothalamus (Leibowitz 1983). Similar reasoning led Schlemmer, Casper, Narasimhachari, and Davis (1979) to suggest using the alpha-adrenergic agonist clonidine as a possible treatment for anorexia nervosa.

Endogenous opioids

It has been speculated that in anorexia nervosa there may be an excessive production of opioids and that an opiate antagonist, such as naloxone, may be a useful treatment (Gillman and Lichtigfeld 1981). Intravenous infusions of naloxone have indeed been administered to patients with anorexia nervosa, with resulting weight gain when the patients were used as their own controls in an open trial (Moore, Mills, and Forster 1981). These authors, however, attributed the weight gain to an antilipolytic effect of naloxone.

Oestrogens

It has been postulated that anorexia nervosa results from a hypersensitivity of the hypothalamus to oestrogens (Young 1975). Treatment with progesterone is suggested on the basis that it antagonizes some of the effects of oestrogen.

The contradictory nature and multiplicity of the hypotheses based on possible alterations in neurotransmission must lead to the conclusion that so far they are unimpressive. The drug treatments proposed as a result of these theoretical models are also disappointing; few controlled trials have been undertaken, and the improvements that have been reported were generally modest in terms of weight gain and alterations in the patients' abnormal attitudes.

TREATMENT AIMED AT PSYCHIATRIC DISORDERS SUPPOSEDLY UNDERLYING ANOREXIA NERVOSA AND BULIMIA NERVOSA (TABLE 8.2)

The arguments in favour of anorexia nervosa being considered a psychiatric illness in its own right have been well presented in a review by Garfinkel and Garner (1982, pp 20–26), a view also shared by the present author (Russell

TABLE 8.2 *Treatment choice on basis of supposed underlying psychiatric disorder*

?Underlying disorder	Eating disorder	Drug choice	Authors
Depression	AN	Amitriptyline	(Mills *et al.* 1973; Needleman and Waber 1976)
	BN	Mianserin	(Sabine *et al.* 1983)
		Imipramine	(Pope *et al.* 1983)
Mania	AN	Lithium	(Barcai 1977; Gross *et al.* 1981)
Epilepsy	BN	Phenytoin	(Green and Rau 1977; Wermuth *et al.* 1977)

Rationale for treatment depending on the supposition that an alternative psychiatric disorder gives rise to the symptoms of anorexia nervosa (AN) or bulimia nervosa (BN).

1970). Nevertheless a case has been made for a relationship between anorexia nervosa and depressive illness. Cantwell, Sturzenberger, Burroughs, Salkin, and Green (1977) observed that depressive symptoms are frequent in anorexic patients, both during the acute phase of their illness, and subsequently when followed up. They found that a high proportion of first degree relatives had a history of affective disorder, an observation also made by Winokur, March, and Mendels (1980) and Hudson, Harrison, Pope, Jonas, and Yurgelun-Todd (1983). There is no doubt about the frequency of depressive symptoms in anorexic patients, more particularly when they are malnourished. Yet these symptoms become diminished as a result of refeeding and weight gain, indicating that to some degree they are secondary to malnutrition (Channon 1982). Finally, it should be remembered that suicide is the commonest cause of death in anorexia nervosa, again supporting a possible link with affective disorder.

For these reasons, treatment with antidepressants has been advocated. They are commonly prescribed in clinical practice when severe or persistent depressive symptoms are encountered in anorexic patients. The important question is whether antidepressant medication can promote a return to a normal pattern of eating and a restoration of body weight, together with a shortening of the illness. There is a remarkable paucity of controlled trials. Mills, Wilson, Eden, and Lines (1973) treated a large number of their patients with tricyclic antidepressants but provided little convincing evidence that this medication led to an improved outcome. The limitations of trials with antidepressant medication in anorexia nervosa are illustrated in a study by Needleman and Waber (1976). Their main criterion of improvement in the six patients given amitriptyline was that of weight gain which did occur to a varying degree (2.8–16.9 kg). They seemed not to have appreciated that benefits result from merely admitting the patient to hospital, assuming that a basic level of nursing care had been provided. Moreover they do not indicate whether a more fundamental improvement occurred in the course of the patients' illness. Other authors' observations suggest that tricyclic antidepressants do not influence the long-term outcome of anorexia nervosa (Morgan and Russell 1975). It is remarkable that a good clinical trial using antidepressant medication in anorexia nervosa has yet to be carried out.

Controlled trials of antidepressants have been conducted in patients suffering from bulimia nervosa. Such an approach is amply justified in view of the frequency with which depressive symptoms are present in bulimic patients (Russell 1979). A double-blind placebo-controlled trial of mianserin in a series of 50 bulimic patients was carried out by Sabine, Yonace, Farrington, Barratt, and Wakeling (1983). Both the mianserin- and the placebo-treated groups showed some gradual improvement during the eight weeks of the trial. But there was no difference between the effects of mianserin and placebo as judged by the Eating Attitudes Test (Garner and Garfinkel 1979), a bulimia rating scale, and measures of anxiety and depression. Nor was there any diminution in the number of days each week that the patients reported overeating, vomiting or purging. The authors concluded that mianserin produced no detectable benefit in their patients, and added the view that bulimia is not a manifestation of an underlying affective disorder.

In contrast, a placebo-controlled trial of imipramine yielded the conclusion that the frequency and intensity of the episodes of overeating were reduced by this tricyclic antidepressant (Pope, Hudson, Jonas, and Yurgelun-Todd 1983). Other reported benefits included a decrease in the patients' preoccupation with food and a relief of depressive symptoms. The authors also concluded that their findings constituted evidence that bulimia represents a form of affective disorder.

In trying to explain the differences between these two trials of antidepressant therapy, two observations seem apposite. In the mianserin trial a larger group of patients were studied who had been referred for treatment to a psychiatric department, and were probably severely afflicted by their symptoms. On the other hand, the subjects in the imipramine trial were likely to have had a milder bulimic disorder, as they were recruited by advertizing in the newsletter of the local Anorexia Nervosa Aid Society and in a Boston newspaper. Another important difference between the two trials is the choice of the antidepressant. The appropriateness of mianserin as an antidepressant treatment of bulimic patients has been questioned by de Buck (1983) who considered it likely to increase appetite and cause weight gain, effects which would distress these patients. Bulimia nervosa is a disorder which often defies treatment. It is therefore important to establish whether antidepressants are effective in relieving the bulimic as well as the depressive symptoms.

Some authors have pushed further the search for a relationship between eating disorders and other psychiatric illnesses, and have consequently tried out more unusual drugs in the treatment of anorexic and bulimic patients. Once again, the possible relationship between anorexia nervosa and affective illnesses suggested the use of lithium carbonate in a pilot study by Barcai (1977) and in a controlled trial by Gross, Ebert, Faden, Goldberg, Nee, and Kaye (1981). The latter study was carried out on 16 anorexic patients equally divided into lithium- and placebo-treated groups. At the same time as receiving their medication the patients took part in a behaviour modification treatment programme. Both groups of patients gained weight over the course of four weeks; the lithium-treated patients gained an average of 6.8 kg and the placebo-treated patients an average of 5.2 kg. It so happened that the randomization procedure resulted in the allocation of somewhat lighter patients to the placebo group than to the lithium-treated group. Nevertheless the weight gains were significantly greater in the lithium-treated group during the third and fourth weeks of treatment. The authors concluded that lithium carbonate may augment weight gain in anorexic patients also treated with behaviour modification. This conclusion appears justified, especially as the authors recognize that their study was a preliminary one which should be followed by further studies of longer duration with larger groups of patients. But a potentially toxic drug, such as lithium carbonate, should not be administered to anorexic patients until it is shown that it shortens the course of the illness.

A more unusual approach, this time to the treatment of bulimic patients, was advocated by Green and Rau (1977) who administered the anticonvulsant phenytoin on the basis of a 'possible neurophysiological element in compulsive eating'. They reported a diminution of bulimic symptoms in a proportion of their patients. In a controlled study, once again only a proportion of the patients showed some improvement with phenytoin as compared with placebo (Wermuth, Davis, Hollister, and Stunkard 1977). The results of treatment with this drug, and the underlying rationale, are not impressive.

THE EMPIRICAL TREATMENT OF ANOREXIA NERVOSA

It may be useful to present the current approach to the treatment of anorexia nervosa which is based principally on clinical observations. This will help clarify the outstanding therapeutic problems that still delay recovery from an anorexic episode, and make it very difficult to predict the outcome in an individual patient.

There are three main areas of clinical disturbance in anorexia nervosa which together constitute a set of diagnostic criteria for the illness (Russell 1970). They also serve as pointers to the patient's therapeutic needs. The first of these disturbances is the considerable loss of weight that is self-induced, mainly through the studied avoidance of carbohydrate-rich foods. Second, there is a specific psychopathology, the core of which is an over-valued idea that fatness is a dreadful state, to be avoided at all cost. Third, a specific endocrine disorder, involving the hypothalamic-pituitary-gonadal axis, gives rise to amenorrhoea in the female at an early stage of the illness. Although there is much overlap in the phases of treatment aimed at these three main disturbances, it is convenient to describe them in turn.

Restoration of body weight

As the loss of weight is due mainly to a reduction in the intake of energy-containing foods, it follows that treatment should be aimed at increasing food intake to the high levels necessary for a positive calorie balance, and thus restore wasted body tissues. It was pointed out long ago that the term 'anorexia nervosa' is a misnomer, for few of these patients have a true loss of appetite (Kay and Leigh 1954). Hence a treatment aimed simply at increasing the patient's hunger and appetite is inappropriate, because she usually exerts a purposive control over her food intake aimed at a reduction of body weight. This basic observation is an important obstacle to drawing comparisons between the reduced food intake in anorexia nervosa and the aphagia induced in animals by hypothalamic lesions or other experimental procedures. There is in fact no satisfactory animal model of anorexia nervosa.

Yet it is a relatively simple matter to restore the patient's weight to a normal level. The treatment does not require any of the more heroic procedures which from time to time have been advocated - tube-feeding, parenteral feeding, or indeed any form of medication. Consistent weight gain, and a return of the patient's weight to its optimum level (usually the weight before the onset of the illness), can be ensured by adhering to a regime of general management. This regime does not rely on any specific measures, nor on any theoretical formulation, but it does depend on carefully implemented policies. The first is to secure the patient's trust and co-operation. This begins when the patient is first seen as an out-patient, and a therapeutic relationship is consolidated over the course of two or three interviews. By then the patient usually accepts the therapist's advice to be admitted to a hospital unit where the nursing staff is experienced in the treatment of anorexic patients. At this stage the second set of policies is implemented. They depend on the application of a range of nursing skills (Russell 1977, 1984). The patient is encouraged to put her trust in the nurses who become responsible for her care, including decisions about meals, food intake and increases in weight. The nurses reassure the patient that the treatment will not make her fat, but they nevertheless aim at restoring her weight to its pre-illness level. In optimum circumstances, the degree of patient supervision that is necessary remains unobtrusive, although the patient should always be confined to the ward and have all her meals in the company of a nurse who ensures that all the food presented is consumed.

The results of this therapeutic regime are nearly always satisfactory in terms of weight gain. Depending on the patient's weight on admission and the optimum weight aimed at, it is usually possible to achieve a weight gain of 15–16 kg in 50 days (Russell 1984). An important lesson to be learned from such a satisfactory response is that therapeutic trials involving a new drug, for example, should take into account these impressive weight gains that can be attained through relatively simple and non-specific measures. Hence, for a new treatment to be of value, it should accomplish more than a simple gain in weight, when this is no problem, as in the course of in-patient treatment. It should bring about a more fundamental change in the patient's ideation and mood, and preferably demonstrate its superiority in maintaining this improvement.

Improvements in the patient's thinking, attitudes, and mood

A gratifying aspect of the hospital management which leads to weight gain is that the patient's mental state often improves simultaneously. Thus her abnormal attitudes to her body weight and size may diminish and her depressed mood lighten (Channon 1982). The patient may continue to progress well after discharge from hospital. There is, however, a substantial proportion of patients whose abnormal attitudes persist, with the consequence that they gradually lose weight after discharge, and may indeed suffer a full relapse (Morgan and Russell 1975).

It must be conceded that the efficacy of treatments aimed at the maintenance of weight outside hospital, and the prevention of relapse, has not so far been established. There is no drug treatment currently available which goes any way towards meeting these objectives. The preferred treatments involve psychotherapeutic approaches, including family therapy (Liebman, Minuchin, and Baker 1974; Minuchin, Baker, Rosman, Liebman, Milman, and Todd 1975; Selvini Palazzoli 1978), and cognitive behavioural treatment (Garner and Bemis 1982). A current MRC-supported therapeutic trial comparing individual supportive psychotherapy and family therapy at the Maudsley Hospital has provided preliminary findings of relevance to the continued co-operation of patients with their treatment (Szmukler, Eisler, Russell, and Dare 1985). It has been shown that about one in four patients fail to become engaged in their treatment after discharge from hospital; they 'drop-out' within the first three months. The drop-out rate is approximately the same for supportive therapy and family therapy, but its occurrence depends mainly on the nature of family interactions. The most useful prediction of 'dropping out' from treatment proved to be that of 'Expressed emotion', a measure devised by Brown, Birley and Wing (1972), and Vaughn and Leff (1976). For example, if the mothers of patients in family therapy expressed frequent critical comments about their daughters, there was a tendency for the family to discontinue treatment. The complexity of such therapeutic issues renders it less likely that pharmacological treatments will provide easy solutions.

Restoration of menstrual cycles

Most patients whose weight is restored to normal spontaneously resume cyclical menstruation after an interval ranging from one to several months. In their case, drug or hormonal treatment is unnecessary. In patients who remain below their optimum weight, there is no point inducing menstruation with hormonal therapy as this does not lead to any benefit in terms of weight gain or improved mental state. There is, however, a place for the treatment of persistent amenorrhoea if, for example, menstruation has not returned in a young woman who has maintained a healthy weight for six months, and is keen for her cycles to re-appear. The endocrinology of anorexia nervosa is an area of study in which clear advances have been made during the past twenty years. As a result of this knowledge, including the role of the negative and positive feedback actions of oestrogen on the hypothalamus, treatment can be given which will induce menstruation and ovulation (Wakeling, DeSouza, and Beardwood 1977). Nillius and Wide (1977) have actually induced ovulatory cycles in anorexic patients by the use of luteinizing hormone releasing hormone (LHRH), even though they were still underweight and showed little spontaneous gonadotrophin or ovarian activity. As already mentioned, such treatment, though of theoretical interest, has little practical justification. In patients who are of normal weight, however, there may be a response to a course of clomiphene citrate which blocks the negative feedback action of oestrogen on the hypothalamus and results in an increased gonadotrophin release. Menstruation will occur if the hypothalamus responds to the subsequent positive feedback action of oestrogen with a surge in the release of LH. If the positive feedback action is absent, it can be mimicked by giving LHRH in a saline infusion 14 days after the commencement of clomiphene treatment, when ovulation will usually be induced (Wakeling and de Souza, quoted by Szmukler 1982).

THE EMPIRICAL TREATMENT OF BULIMIA NERVOSA

Severely bulimic patients are less responsive to treatment than patients with anorexia nervosa. This is sometimes because the bulimic symptoms represent a chronic stage in the patient's anorexia nervosa (Russell 1979). The illness has three principal components which correspond to its diagnostic criteria, and provide guidance to empirical methods of treatment. First there is a preoccupation with food which leads to episodic gorging. Secondly, the patient endeavours to avoid the 'fattening' effects of the ingested food by adopting one or more devices among which self-induced vomiting and self-purging are commonest. Thirdly, there is a psychopathology identical with that of anorexia nervosa, whereby the patient dreads becoming fat, and tries to maintain her weight below a threshold value which she considers to be her 'ideal weight'.

Although self-induced vomiting and purging are the most dramatic features of the illness, these abnormal behaviours tend to disappear once the patient is able to control her food intake and avoid the episodes of gorging. Current methods of treatment are therefore aimed at helping the patient adopt a more normal pattern of eating. One method is based on a cognitive behavioural regime involving a structured regulation of meals, increasing control over eating, exposure to foods otherwise avoided, and improving attitudes to body weight (Fairburn 1981). Another method depends on encouraging the patient to accept the higher weight levels which she dreads. This usually requires admission to hospital, when the patient is prevailed upon to let her weight rise to a level corresponding with her 'healthy' weight, namely, the weight which she maintained before the onset of her illness (Russell 1984). There is also a place for individual psychotherapy or family therapy. It must be admitted, however, that so far there is no clear evidence as to the efficacy of any of these treatments.

The possible role of drug treatment of bulimic patients has already been discussed, with the greatest promise held out for the tricyclic antidepressant imipramine (Pope *et al.* 1983).

An experimental approach to the treatment of bulimia nervosa

An essential clinical feature of bulimia nervosa is the patient's preoccupation with food. Although she will often deny that her abnormal sensations are equivalent to heightened hunger, she will usually accept that they amount to cravings for food. Thus, bulimic patients present fewer problems than anorexic patients in conceptualizing their illness in terms of a disturbance in the normal regulation of food intake. This observation has led our group at the Institute of Psychiatry to investigate the response of bulimic patients to anorexic drugs under laboratory conditions.

The two drugs studied were methylamphetamine and fenfluramine. The actions of amphetamine and fenfluramine have been contrasted in animals and man by Blundell, Latham, Moniz, McArthur, and Rogers (1979). They have one action in common, namely, the capacity to reduce the weight of food consumed. They possess, however, contrasting actions on brain neurotransmitters, amphetamine acting mainly on dopamine and noradrenaline, fenfluramine mainly on 5-hydroxytryptamine. The two drugs operate through separate anatomical systems and differ in the way they influence the process of eating. In man (as in animals) amphetamine delays the onset of eating, whereas fenfluramine slows the rate of eating. This suggests an effect of amphetamine on hunger and an action of fenfluramine on satiety (Blundell, Latham, and Leshem 1976). In man the anorexic effects of fenfluramine are more enduring than those of amphetamine. This information was helpful in designing experiments to test the differential effects of the two drugs on bulimic patients.

Methylamphetamine

In the first set of experiments with methylamphetamine, eight bulimic patients were each tested on two occasions (Ong, Checkley, and Russell 1983). After an overnight fast, they were given an intravenous infusion of methylamphetamine (15 mg/75 kg) or placebo. The order of administration was randomized and the observers were 'blind' to the substance administered. Self-rated visual analogue scales were completed at 15-minute intervals for sensations such as 'not hungry – extremely hungry'. Observer ratings were made on mood and increased activation (alertness, excitement, pressure of talk). Each patient was left alone 120 minutes after the methylamphetamine or placebo injection with a large supply of food of her own choice. The caloric content of the food consumed during 30 minutes was measured for each test.

Figure 8.1 shows the means of the self-ratings for hunger in eight patients, plotted for the two hours after the injection of methylamphetamine and placebo. The difference in the areas below the two curves was statistically significant ($t = 3.37; p < 0.02$). The action of methylamphetamine in reducing hunger lasted approximately two hours. There was also a significant reduction in the caloric consumption after methylamphetamine (224±111 cal) compared with placebo (943±222 cal), ($t = 3.0; p < 0.02$). Moreover, whereas four out of eight patients succumbed to a bulimic episode after placebo, none did so after methylamphetamine. Thus methylamphetamine reduced symptoms characteristic of bulimia, but whether this was due to a reduction in hunger or the stimulant and euphoriant effects of the drug could not be established.

Fenfluramine

In a second study, 15 bulimic patients were given either fenfluramine (60 mg orally) or placebo under double-blind and randomized conditions (Robinson,

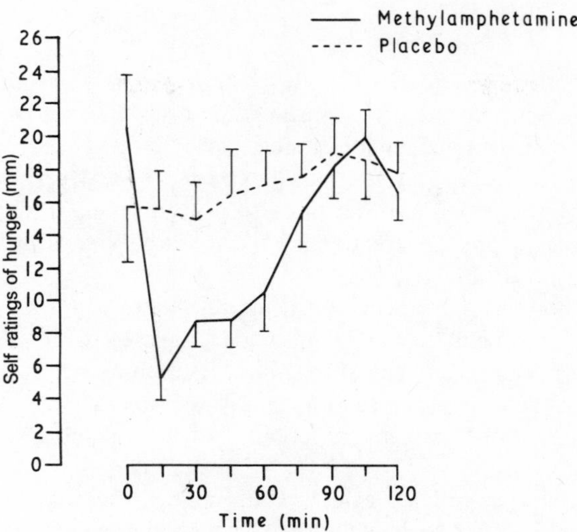

FIG. 8.1 Mean self-ratings of hunger in 8 patients with bulimia nervosa, before and after the injection of methylamphetamine (15 mg/75 kg) (–) or placebo (- - - -). Reproduced with permission of Ong *et al.* (1983) and the *Br. J. of Psychiat.*

Checkley, and Russell 1985). The patients were again left alone with a liberal supply of food two hours after taking fenfluramine or placebo. In this study, they were additionally asked to record after leaving the laboratory and for the next five days episodes of overeating (often called 'binges' by the patients). This provided a measure of bulimic episodes under natural conditions during the days following the administration of the tablets.

The results showed that during the course of the two hours' experiment the patients rated themselves on a visual analogue scale as experiencing less hunger after taking fenfluramine, whereas after placebo there was an increase in their feeling of hunger. Fenfluramine was followed by a smaller calorie intake (214±38 cal) during the test meal than after placebo (494±75 cal; $p < 0.01$). Four out of 15 patients experienced a bulimic episode after placebo; none did after fenfluramine. Moreover, there were significantly fewer patients reporting at least one episode of bulimia in the 12 hours after fenfluramine compared with placebo ($\text{Chi}^2 = 4.0$, $p < 0.05$). The patients on fenfluramine commonly reported drowsiness as a side-effect. In view of the enhanced alertness and euphoria resulting from methylamphetamine in the first study, it is unlikely that changes in alertness and mood accounted for the suppression of eating after fenfluramine.

Both drugs are capable of suppressing bulimic episodes. Fenfluramine is the more promising drug from the therapeutic point of view, because the stimulant effects of methylamphetamine would carry the danger of dependence and drug-induced psychosis. It must be stressed, however, that the aim of treatment should remain one of enhancing total food intake, even though the patient may obtain relief from a drug which acts by suppressing her distressing bulimic symptoms.

SUMMARY

1. Theoretical models of the genesis of anorexia nervosa based on altered neurotransmitter activity are disappointing: the drug therapies so far predicted from them have failed to yield clinically impressive results.

2. Drug treatment aimed at altering the course of an affective illness supposedly underlying the eating disorder has also proved disappointing. A well conducted trial of antidepressant treatment has not yet been carried out in anorexia nervosa.

3. There is no drug treatment which approximates to the impressive short-term results of a general hospital and nursing management of patients with anorexia nervosa. The assessment of long-term treatments awaits the outcome of trials of psychological therapies such as family therapy. There is a limited place for drug and hormonal treatment in patients in whom amenorrhoea may persist in spite of weight gain.

4. There is some preliminary evidence that imipramine may be of benefit to bulimic patients. Further trials including other antidepressants in patients with more severe forms of bulimia nervosa are awaited.

5. Preliminary results with fenfluramine suggest that bulimic symptoms may be suppressed by single doses of anorexic drugs. A controlled trial of long-term treatment with fenfluramine in bulima nervosa is indicated.

References

Anand, B. K. (1961). Nervous regulation of food intake. *Physiol. Rev.* **41**, 677.

—— and Brobeck, J. R. (1951). Localization of a 'feeding centre' in the hypothalamus of the rat. *Proc. Soc. Exp. Biol. Med. (N.Y.)* **77**, 323.

Barcai, A. (1977). Lithium in adult anorexia nervosa: a pilot report on two patients. *Acta psychiat. scand.* **55**, 97–101.

Barry, V. C. and Klawans, H. L. (1976). On the role of dopamine in the pathophysiology of anorexia nervosa. *J. Neur. Transm.* **38**, 107–22.

Blundell, J. E., Latham, C. J., and Leshem, M. B. (1976). Differences between the anorexic actions of amphetamine and fenfluramine – possible effects on hunger and satiety. *J. Pharm. Pharmac.* **28**, 471–7.

—— —— Moniz, E., McArthur, R. A., and Rogers, P. J. (1979). Structural analysis of the actions of amphetamine and fenfluramine on food intake and feeding behaviour in animals and in man. *Current Medical Research and Opinion* **6**, Suppl. 1, 34–54.

Brown, G. W., Birley, J. L. T., and Wing, J. K. (1972). Influence of family life on the course of schizophrenic disorders: a replication. *Br. J. Psychiat.* **121**, 241–58.

Cantwell. D. P., Sturzenberger, S., Burroughs, J. Salkin, B., and Green, J. K. (1977). Anorexia nervosa: an affective disorder? *Arch. Gen. Psychiat.* **34**, 1087–93.

Channon, S. (1982). A study of correlates of weight gain during the treatment of anorexia nervosa. M. Phil. dissertation; University of London.

Coscina, D. V. (1977). Brain amines in hypothalamic obesity. In *Anorexia nervosa* (ed. R. A. Vigersky) pp. 97–107. Raven Press, New York.

De Buck, R. (1983). Discussion of 'Bulimia nervosa: a placebo controlled double-blind therapeutic trial of mianserin'. *Br. J. Pharmac.* **15**, 201S–202S.

Dally, P. J. (1967). Anorexia nervosa – long-term follow-up and effects of treatment. *J. Psychosom. Res.* **11**, 151.

—— and Sargant, W. (1960). A new treatment of anorexia nervosa. *Br. Med. J.* **1**, 1770.

—— —— (1966). Treatment and outcome of anorexia nervosa. *Br. Med. J.* **2**, 793.

Fairburn, C. (1981). A cognitive behavioural approach to the treatment of bulimia. *Psychol. Med.* **11**, 707–11.

Garfinkel, P. E. and Garner, D. M. (1982). *Anorexia nervosa: a multidimensional perspective.* Brunner Mazel, New York.

Garner, D. M. and Bemis, K. M. (1982). A cognitive-behavioural approach to anorexia nervosa. *Cogn. Ther. Res.* **6**, 123–50.

—— and Garfinkel, P. E. (1979). The eating attitudes test; an index of the symptoms of anorexia nervosa. *Psychol. Med.* **9**, 273–9.

Gillman, M. A. and Lichtigfeld, F. J. (1981). Naloxone in anorexia nervosa: role of the opiate system. *J. R. Soc. Med.* **74**, 631.

Goldberg, S. C., Halmi, K. A., Eckert, E. D., Casper, R. C., and Davis, J. M. (1979). Cyproheptadine in anorexia nervosa. *Br. J. Psychiat.* **134**, 67–70.

Green, R. S. and Rau, J. H. (1977). The use of diphenylhydantoin in compulsive eating disorders: further studies. In *Anorexia nervosa* (ed. R. A. Vigersky) pp. 377–82. Raven Press, New York.

Gross, H. A., Ebert, M. H., Faden, V. B., Godberg, S. C., Nee, L. E., and Kaye, W. H. (1981). A double-blind controlled trial of lithium carbonate in primary anorexia nervosa. *J. Clin. Psychopharmac.* **1**, 6, 376–81.

Harrower, A. D. B., Yap, P. L., Nairn, I. M., Walton, H. J., Strong, J. A., and Craig, A. (1977). Growth hormone, insulin and prolactin secretion in anorexia nervosa and obesity during bromocriptine treatment. *Br. Med. J.* **2**, 156–9.

Hudson, J. I., Harrison, G., Pope, H. G., Jonas, J. M., and Yurgelun-Todd, D. (1983). Family history study of anorexia nervosa and bulimia. *Br. J. Psychiat.* **142**, 133–8.

Johanson, A. J. and Knorr, N. J. (1977). L-Dopa as treatment for anorexia nervosa. In *Anorexia Nervosa* (ed. R. A. Vigersky) pp. 363–72. Raven Press, New York.

Kay, D. W. K. and Leigh, D. (1954). The natural history, treatment and prognosis of anorexia nervosa, based on a study of 38 patients. *J. Ment. Sci.* **100**, 411.

Leibowitz, S. F. (1983). Hypothalamic catecholamine systems controlling eating behavior: a potential model for anorexia nervosa. In *Anorexia nervosa; recent developments in research* (eds. P. L. Darby, P. E. Garfinkel, D. M. Garner and D. V. Coscina) Vol. 3, Neurology and neurobiology, pp. 221–9. Alan R. Liss, New York.

Liebman, R., Minuchin, S., and Baker, L. (1974). An integrated treatment program for anorexia nervosa. *Am. J. Psychiat.* **131**, 432–6.

Mawson, A. R. (1974). Anorexia nervosa and the regulation of intake: a review. *Psychol. Med.* **4**, 289–308.

Mills, I. H., Wilson, R. J., Eden, M. A. M., and Lines, J. G. (1973). Endocrine and social factors in self-starvation amenorrhoea. In *Symposium – Anorexia Nervosa and Obesity* (ed. R. F. Robertson) pp. 31–43. Royal College of Physicians, Edinburgh.

Minuchin, S., Baker, L., Rosman, B. L., Liebman, R., Milman, L., and Todd, T.C. (1975). A conceptual model of psychosomatic illness in children. *Arch. Gen. Psychiat.* **32**, 1031–8.

Moore, R., Mills, I. H., and Forster, A. (1981). Naloxone in the treatment of anorexia nervosa: effect on weight gain and lipolysis. *J. R. Soc. Med.* **74**, 129–31.

Morgan, H. G. and Russell, G. F. M. (1975). Value of family background and clinical features as predictors of long-term outcome in anorexia nervosa: four year follow-up study of 41 patients. *Psychol. Med.* **5**, 355–71.

Needleman, H. L. and Waber, D. (1976). Amitriptyline therapy in patients with anorexia nervosa. *Lancet* **ii**, 580.

Nillius, S. J. and Wide, L. (1977). The pituitary responsiveness to acute and chronic administration of gonadotropin-releasing hormone in acute and recovery stages of anorexia nervosa. In *Anorexia nervosa* (ed. R. A. Vigersky) pp. 225–41. Raven Press, New York.

Ong. Y. L., Checkley, S. A., and Russell, G. F. M. (1983). Suppression of bulimic symptoms with methylamphetamine. *Br. J. Psychiat.* **143**, 288–93.

Pope, H. G., Hudson, J. I., Jonas, J. M. and Yurgelun-Todd, D. (1983). Bulimia treated with imipramine: a placebo-controlled, double-blind study. *Am. J. Psychiat.* **140**:**5**, 554–8.

Redmond, D. E., Swann, A. and Heninger, G. R. (1976). Phenoxybenzamine in anorexia nervosa (letter). *Lancet* **ii**, 307.

—— Huang, Y. H., Baulu, J., Snyder, D. R., and Mass, J. W. (1977). Norepinephrine and satiety in monkeys, in anorexia nervosa. In *Anorexia nervosa* (ed. R. A. Vigersky) pp. 81–96. Raven Press, New York.

Robinson, P. H., Checkley, S. A., and Russell, G. F. M. (1985). Suppression of symptoms of bulimia nervosa with fenfluramine. *Br. J. Psychiat.* **146**, 169–76.

Russell, G. F. M. (1970). Anorexia nervosa: its identity as an illness and its treatment. In *Modern trends in psychological medicine* (ed. J. Harding Price) Vol. 2, pp. 131–64. Butterworths, Norwich.

—— (1977). General management of anorexia nervosa and difficulties in assessing the efficacy of treatment. In *Anorexia Nervosa* (ed. R. A. Vigersky), pp. 277–89. Raven Press, New York.

—— (1979). Bulimia nervosa: an ominous variant of anorexia nervosa. *Psychol. Med.* **9**, 429–48.

—— (1984). Anorexia nervosa and bulimia nervosa. In *Handbook of Psychiatry 4: The Neuroses and Personality Disorders* (eds. G. F. M. Russell and L. Hersov), pp. 285–98. Cambridge University Press, Cambridge.

Sabine, E. J., Yonace, A., Farrington, A. J., Barratt, K. H., and Wakeling, A. (1983). Bulimia nervosa: a placebo controlled double-blind therapeutic trial of mianserin. *Br. J. Clin. Pharmac.* **15**, 195S–202S.

Schlemmer, R. F., Casper, R. C., Narasimhachari, N., and Davis, J. M. (1979). Clonidine induced hyperphagia and weight gain in monkeys. *Psychopharmacologica* **61**, 233–4.

Selvini Palazzoli, M. (1978). *Self-starvation: from individual to family therapy in the treatment of anorexia nervosa.* (Translated by A. Pomerans) pp. 193–201. Aronson, New York.

Silverstone, T. and Turner, P. (1974). *Drug treatment in psychiatry.* p. 185. Routledge and Kegan Paul, London.

Szmukler, G. I. (1982). Drug treatment of anorexic states. In *Drugs and appetite* (ed. J. T. Silverstone) pp. 159–81. Academic Press, London.

—— Eisler, I., Russell, G. F. M., and Dare, C. (1985). Anorexia nervosa, parental 'expressed emotion' and dropping out of treatment. *Br. J. Psychiat.* (In press).

Ungerstedt, U. (1971*a*). Stereotaxic mapping of the monoamine pathways in the rat brain. *Acta physiol. scand. Suppl.* **367**, 1–48.

—— (1971*b*). Adipsia and aphagia after 6-hydroxydopamine induced degeneration of the nigro-striatal dopamine system. *Acta physiol scand. Suppl.* **367**, 95–122.

Vaughn, C. E. and Leff, J. P. (1976). The measurement of expressed emotion in the families of psychiatric patients. *Br. J. Soc. Clin. Psychol.* **15**, 157–65.

Vandereycken, W. (1984). Neuroleptics in the short-term treatment of anorexia nervosa. A double-blind placebo-controlled study with sulpiride. *Br. J. Psychiat.* **144**, 288–92.

—— and Pierloot, R. (1982). Pimozide combined with behaviour therapy in the short-term treatment of anorexia nervosa. *Acta psychiat. scand.* **66**, 445–50.

Vigersky, R. A. and Loriaux, D. L. (1977). Anorexia nervosa as a model of hypothalamic dysfunction. In *Anorexia nervosa* (ed. R. A. Vigersky) pp. 109–21. Raven Press, New York.

Wakeling, A., DeSouza, V. A., and Beardwood, C. J. (1977). Assessment of the negative and positive feedback effects of administered oestrogen on gonadotrophin release in patients with anorexia nervosa. *Psychol. Med.* **7**, 397–405.

Wermuth, B. M., Davis, K. L., Hollister, L. E., and Stunkard, A. J. (1977). Phenytoin treatment of the binge-eating syndrome. *Am. J. Psychiat.* **134(11)**, 1249–53.

Winokur, A., March, V., and Mendels, J. (1980). Primary affective disorder in relatives of patients with anorexia nervosa. *Am. J. Psychiat.* **137**, 695–8.

Young, J. K. (1975). A possible neuroendocrine basis of two clinical syndromes: anorexia nervosa and the Kleine-Levin syndrome. *Physiol. psychol.* **3(4)**, 322–30.

9

Vitamin deficiencies: a factor in senile dementia?

K. O. CHUNG-A-ON, D. E. THOMAS, S. F. TIDMARSH, J. W. T. DICKERSON, E. A. SWEENEY, AND D. M. SHAW*

Dementia occurs in about 2.4 per cent of persons aged 65 to 69 years but in 22 per cent of those ages 80 years and over (Kay Beamish, and Roth 1964; Kay, Bergman, and Foster 1970). The importance of these observations is emphasized by the increasing proportion of the population reaching old age, and the increasing demand that they make on health care facilities. The causes of dementia are unknown. A prominent feature is neuronal loss, and it had been suggested that other changes, such as a reduction in the activity of choline acetyltransferase, a marker of the presynaptic cholinergic system, are secondary to this. There may be a deficiency in serotoninergic and other transmission systems in this condition. Preliminary studies by Lehmann (1979) suggested that a proportion of old people may not absorb tryptophan normally, and that this is associated with dementia. Shaw, Tidmarsh, Sweeney, Williams, Karajgi, Elameer, and Twining (1981) reported significantly lower levels of tryptophan, the precursor of brain serotonin, in the plasma of patients with senile dementia when compared with values in elderly controls. This finding is of considerable potential interest in relation to diet because primary dietary lack, or perhaps more likely, a failure to absorb certain nutrients, might contribute to the aetiology of the disease. This study showed also that the ratios of the concentrations of tryptophan in plasma (total and non-protein bound), to five neutral amino acids were significantly lower in patients with senile dementia than in the normal elderly.

The entry of tryptophan into the brain is not controlled solely by the availability of this amino acid, but is modified by the ratio of the concentration of tryptophan to the sum of these other neutral amino acids (phenylalanine, tyrosine, leucine, isoleucine, and valine), which compete for the same transport mechanism (Fernstrom, Madras, Munro and Wurtman 1974). Thus a low ratio of tryptophan to these amino acids will tend to result in low levels of tryptophan in the brain. Other factors modifying the concentration and binding of tryptophan in plasma are albumin, and non-esterified fatty acids (Curzon, Friedel and Knott 1973), and insulin modifies the ratio of tryptophan to the five neutral amino acids (Dickerson and Pao 1975). Drugs also alter the 'disposition' of tryptophan.

If there is a degree of malabsorption in some old people, as suggested by Lehmann (1979), it is unlikely that it affects the absorption of tryptophan specifically. Indeed, it is tempting to suggest that malabsorption is one factor

* All correspondence to D. M. Shaw.

contributing to the finding of low levels of vitamin C (Schorah 1979) and folate (Dickerson 1978) in many old people. These, and a number of other vitamins, play important roles in brain function.

There is abundant evidence in the literature that there are groups within the elderly population that are particularly vulnerable to nutritional deficiency (Exton-Smith 1978). In many studies (e.g. Vir and Love 1979) dietary evidence of deficiency has been supported by biochemical assessments. However, in no study known to us has there been an attempt to relate dietary intake with nutrition nor with factors which may influence synthesis of serotonin in the brain in patients with senile dementia. This paper reports the results of such a study.

METHODS

Selection of patients

In-patients at the Royal Hamadryad and Whitchurch Hospitals, Cardiff, were diagnosed by the psychogeriatricians in charge (Dr. E. Sweeney and Dr. I. Wilson) as severely demented, on the basis of history, clinical characteristics, performance on the mental test and higher neurological function scales. It was not possible to obtain sufficient drug-free patients. Those on the minimum number of drugs were selected, and none was accepted if the drug they were taking might interfere with vitamin metabolism (e.g. anticonvulsants and vitamin D; Stamp, Round, Rowe and Haddad 1972), or with intestinal absorption.

Selection of controls

All were healthy, ambulant individuals over 65 years of age who came following our requests for volunteers made via local advertising. None was taking drugs at the time of the investigation, and all completed the mental test and higher neurological function scales successfully (Hodkinson 1973) thus eliminating candidates with senile memory loss (Table 9.1).

Physical illness

No patient or control was accepted who had evidence of major disease on physical examination, or in the routine laboratory tests.

TABLE 9.1 *Details of patients with senile dementia and healthy controls*

	Sex	Number	Age (years)	
			Mean	Range
Senile dements	M	9	75.4	66-83
	F	20	80.1	68-89
Controls	M	16	73.9	69-82
	F	19	74.7	66-85

Ethical approval

The study was approved by the local and Area Ethical Committees, and the control subjects gave informed consent to participate in the study. Consent for the patients, who had advanced senile dementia, was sought from the nearest relative or, if no such relative was available, from the patient's general practitioner.

Dietary assessments

Patients

Each item of food and drink actually consumed by the patients was weighed and recorded by a dietician on three consecutive days. The nutrient intake was calculated on the Welsh National School of Medicine computer using McCance and Widdowson's food tables (Paul and Southgate 1979). The dietary data presented in this paper are the mean intake for the 3-day period.

Controls

Three-day weighed dietary measurements were recorded by the subjects. Scales were provided and the procedure of weighing their food was explained to them in detail. Examples were given and several test meals were tried. Dietary intake was calculated over the three days and, during that period, the subjects were visited to ensure that the procedure was being followed correctly.

Samples

On the third day of monitoring the diet, an attempt was made to obtain a urine sample over a 4-h period for determination of *N*-methylnicotinamide (NMN) (Goldsmith and Miller 1967) and creatinine (Bonsnes and Taussky 1945). A fasting sample of blood was taken by venepuncture with stasis on the fourth day, and was suitably partitioned for subsequent biochemical analysis.

Biochemical determination

Plasma and leucocyte ascorbic acid levels were determined by the method of Denson and Bowers (1961) which measures total ascorbic acid, i.e. ascorbic acid, dehydroascorbic acid and diketogulonic acid. Thiamine, riboflavine, and pyridoxine status were estimated by appropriate enzyme activation (Bayoumi and Rosalki 1976). Folate levels in red blood cells and whole blood, and insulin in plasma were determined using radioassay kits (Amersham International Ltd). Total plasma tryptophan was assayed by the method of Denckla and Dewey (1967) and free tryptophan was determined at 37 °C by the ultra-filtration method described by Riley and Shaw (1981). The concentration of albumin in plasma was estimated by rocket electrophoresis (Laurell 1966) and that of non-esterified fatty acids by the technique of Duncombe (1964). The content of ascorbic acid in raw, cooked and served potato was estimated by the 2, 6-dichlorophenolindophenol titration method, as described by the Association of Vitamin Chemists (1966).

RESULTS

The energy intakes of both male and female patients with senile dementia were significantly lower that those of corresponding controls (Table 9.2). None of the men with senile dementia, and only a small proportion of the women, ate the recommended daily amount (RDA) (DHSS 1979). It is to be noted, however, that 50 per cent of the control subjects of both sexes also consumed less than the RDA. The intake of protein by the male patients was similar to that of controls, whereas the intake by female patients was significantly lower than that of the controls. There was no difference in intake of tryptophan by patients and controls of either sex.

TABLE 9.2 *Mean daily intake of energy, protein and tryptophan by patients with senile dementia and control subjects*

	Group	Sex	Intake	RDA*	Percentage below RDA
Energy (kcals)	Senile dements	M	1536 (1177–1699)	2275	100‡
	Controls	M	2157 (1430–2931)		50
	Senile dements	F	1474 (751–2159)	1790	85‡
	Controls	F	1725 (801–2999)		53
Protein (g)	Senile dements	M	63 (±8)	57	22
	Controls	M	68(±5)		6
	Senile dements	F	55† (±11)	44.5	10
	Controls	F	64 (±15)		10
Tryptophan (mg)	Senile dements	M	796 (±111)		
	Controls	M	936 (±214)		
	Senile dements	F	726 (±183)		
	Controls	F	783 (±203)		

Values are means and ranges or means ± standard deviation

* Mean recommended daily allowance for 65 and over age group for men and 55 and over for women (DHSS 1979).

‡ Chi^2 $p<0.001$.

† Student's t-test $p<0.05$.

The mean calculated intake of ascorbic acid exceeded the RDA in both the patients and control subjects (Table 9.3). However, 41 per cent of the patients and 26 per cent of the controls received less than a calculated 30 mg per day. The intake of vitamin C was likely to be lower than estimated because most of the vitamin comes from potato in the hospital diet. The loss of ascorbic acid in hospital from the raw vegetable, through preparation to service, left only 8 per cent of the original vitamin content (Table 9.4).

Few patients were receiving less than the RDA of nicotinic acid and calcium. In contrast, the calculated folate intake of almost all subjects was less than the RDA, and all received less than the RDA of vitamin D. A significantly greater number of patients than controls received less than the RDA of pyridoxine and of iron.

Table 9.5 shows results of the estimates of vitamin status in the two groups. The range of individual values was wide for the different measurements; however, only in the case of thiamine did the mean value indicate deficiency, and there was no significant difference in the proportion of dements and controls that were deficient. In contrast, significantly more dements than controls had lower plasma ascorbic acid and serum folic acid concentrations.

The mean concentrations of total tryptophan, of bound tryptophan, non-esterified fatty acids (NEFA), and albumin in plasma were significantly lower in patients of both sexes (Table 9.6). Satisfactory urine samples for determination of NMN were obtained only in the female patients, and these values were compared with matched female controls. The ratio of NMN to creatinine was significantly lower in the female patients than in controls. Mean free tryptophan and insulin values were similar in patients and controls.

In the 19 patients and 17 controls, mean NMN values were highly significantly different, those of the patients being much lower.

Pearson correlations were calculated between various pairs of measurements. These showed that in controls, but not in patients, plasma concentrations of total and bound tryptophan were significantly correlated with dietary intake of tryptophan ($r = +0.363$ and $+0.379$ respectively), and plasma total, and bound tryptophan levels were negatively correlated with the plasma concentration of NEFA ($r = -0.464$ and -0.577 respectively). There were no correlations. in either group between plasma albumin and the various tryptophan concentrations. There was a similar absence of correlations in both groups between plasma insulin and tryptophan levels. In both groups, as was to be expected, there was a highly significant correlation between plasma total and bound tryptophan concentrations ($r = 0.970$ for controls and $r = 0.981$ for patients). There was a significant correlation between plasma total and free tryptophan only in the patients ($r = 0.4930$).

DISCUSSION

It has been suggested that undernutrition is common and obesity rarely found in patients in psychogeriatric wards. Our results agree with those of others (Asplund, Normonk, and Patterson 1981) in suggesting that dietary energy

TABLE 9.3 *Nutrient intake of patients with senile dementia and control subjects*

Nutrient	Group	Mean intake (range)	RDA*	Percentage below RDA
Ascorbic acid (mg)	Dements	34.5 (17.8–63.2)	30	41
	Controls	43.1 (8.3–99.0)		26
Thiamine (mg)	Dements	1.05 (0.5–1.77)	Men 0.9	28
	Controls	1.05 (0.42–1.60)	Women 0.7	17
Riboflavine (mg)	Dements	1.70 (1.0–2.9)	Men 1.6	28
	Controls	1.67 (0.73–3.08)	Women 1.3	43
Pyridoxine (mg)	Dements	0.94 (0.46–1.43)	1–2	69†
	Controls	1.09 (0.47–1.80)		42
Nicotinic acid (mg)	Dements	23.1 (11.8–38.2)	Men 18	7
	Controls	30.7 (11.6–52.0)	Women 15	6
Folic acid (μg)	Dements	136 (70–193)	200	100
	Controls	145 (80–267)		88
Calcium (mg)	Dements	779 (473–1130)	500	4
	Controls	927 (429–1698)		9
Iron (mg)	Dements	8.7 (4.3–17.9)	10	83‡
	Controls	10.3 (3.9–15.9)		46
Vitamin D (μg)	Dements	2.24 (0.44–5.59)	10	100
	Controls	2.27 (0.29–8.01)		100

* Mean recommended daily allowance for 65 and over age group for men, and 55 and over for women (DHSS 1979).

† Chi² p value < 0.05.

‡ Chi² p value < 0.01.

TABLE 9.4 *Percentage loss of ascorbic acid by potato at Whitchurch Hospital between preparation and serving*

Time	State of potato	Ascorbic acid g per 100g potato	Percentage loss
9.00 am	Raw	10.00	
11.15 am	Immediately after cooking	5.11	49
12.30 pm	When served	0.89	92

TABLE 9.5 *Vitamin status of patients with senile dementia and control subjects*

Vitamin	Subjects	(No.)	Mean	Range	Limit of normal	Percentage with low status
Ascorbic acid Plasma (mg/100ml)	Dements Controls	(28) (35)	0.45 0.78	(0–1.0) (0.15–1.6)	>0.4	57* 23
Buffy coat ascorbic acid (μg/10⁸ cells)	Dements Controls	(29) (35)	34.0 44.2	(4.1–72.1) (0.96–76.6)	>18	4 6
Thiamine Erythrocyte trans-ketolase activation (%)	Dements Controls	(22) (34)	31 30	(0–246) (0–319)	<15	36 55
Riboflavine Erythrocyte gluta-thione reductase activation (%)	Dements Controls	(25) (31)	30 31	(0–68) (9–100)	<76	0 4
Pyridoxine Erythrocyte aspartate amino transferase activation (%)	Dements Controls	(26) (34)	65 59	(0–374) (0–110)	<130	8 0
Folic acid Serum (mg/ml)	Dements Controls	(29) (35)	5.1 6.7	(0.7–12.1) (0.3–15)	>2	23† 3
Erythrocyte (ng/ml)	Dements Controls	(23) (25)	187 293	(100–434) (79–721)	>100	4 8
Nicotinic acid *N*-methylnico-tinamide excretion (mg/g creatinine)	Dements Controls	(20) (30)	10.9 14.8	(4.9–25.5) (2.5–34.3)	3–6	0 3

Value significantly different by Chi^2 analysis shown * when $p < 0.01$ and † when $p < 0.02$.

TABLE 9.6 *Concentrations of various constituents in blood or urine of senile dements and control subjects*

	Sex	Total tryptophan (nmol ml^{-1})	Plasma free tryptophan (nmol ml^{-1})	Plasma bound trytophan (nmol ml^{-1})	Non-esterified fatty acids (nmol ml^{-1})	Albumin (mg ml^{-1})	Insulin (IU)	Urinary *N*-methyl nicotinamide/creatinine (mg/g creatinine)
Dements	M	49.7†	12.2	37.5‡	0.34*	37.0‡	19.2	
		±14.9	±2.5	±13.6	±0.20	±5.5	±6.5	
	No.	9	9	9	8	8	9	
Controls	M	63.6	11.5	52.1	0.53	42.7	21.9	
		±11.1	±2.5	±11.1	±0.23	±4.0	±8.7	
	No.	16	16	16	16	16	14	
Dements	F	48.9§	11.6	37.3§	0.40‡	37.5†	23.4	11.6‡
		±9.0	±2.2	±8.3	±0.23	±4.5	±9.1	±5.0
	No.	20	20	20	20	20	18	16
Controls	F	59.5	11.1	48.3	0.59	41.5	22.8	19.0
		±5.5	±1.8	±5.3	±0.19	±5.1	±9.6	±7.3
	No.	19	19	19	19	19	17	16

Values are means ± SD for number of subjects shown

Values for senile dements that are significantly different from those of controls shown * when $p < 0.05$, † when $p < 0.02$, ‡ when $p < 0.01$, and § when $p < 0.001$.

deficiency may be a contributing factor. Some patients with dementia expend large amounts of energy in constant activity during the day. Patients also appear to want more food than they are given, as they clear their plates and often steal food. It is therefore likely that the energy requirement of this particular group of patients is considerably higher than that of their healthy counterparts.

The intakes an levels in the body of vitamins give cause for concern. Although the mean documented intake of ascorbic acid by both patients and controls exceeded the RDA of 30 mg per day, there must be some doubt about the accuracy of this value due to losses of this vitamin in potato cooked and served in the hospital, as discussed above. Estimates of intakes of the vitamin probably should be reduced by about 50 per cent.

That this is so is supported by the finding that 57 per cent of patients and 23 per cent of controls had plasma ascorbic acid concentrations below the accepted lower limit of normal. According to Loh and Wilson (1971), plasma levels are a more reliable index of ascorbic acid status than tissue levels as represented by 'buffy coat' concentrations. For this measure, only a few subjects (10 per cent of patients with senile dementia and 6 per cent of controls) had values lower than the 18 mg/10^8 as the lower limit of normal suggested by Schorah (1982).

The results are in agreement with those of other psychogeriatric patients (Morgan and Hullin 1982) and of the healthy elderly South Wales population (Burr, Elwood, Hall, Hurley, and Hughes 1974).

The intakes of vitamin D were particularly low in patients, and this is a cause for concern, as they do not go out of doors, and therefore rely solely on dietary vitamin D. Poor intake of vitamin D and of calcium (which was also below the RDA) may make these individuals prone to a possible development of osteomalacia, a condition common in the elderly anyway (Hodkinson, Rand, Station, and Morgan 1973).

Certain vitamins of the B group are of particular relevance to brain function. Thiamine deficiency has been proposed as a cause of confusion in the elderly (A. N. Exton-Smith, personal communication; Older and Dickerson 1982). A fifth of all our elderly subjects were receiving less than the RDA of this vitamin and 55 per cent of the controls and 36 per cent of the patients with senile dementia showed biochemical evidence of thiamine deficiency as judged by elevation of the percentage activation of transketolase (Dreyfus 1962). Elev ted values are seen frequently in geriatric patients (Older and Dickerson 1982; Katakity, Webb, and Dickerson 1983). A number of factors may contribute to thiamine deficiency in the elderly, including destruction of thiamine in cooking, in the stomach by antacids (Dickerson 1978), and by malabsorption.

The intake of folate acid as estimated from the values in food tables were lower than the RDA of 100 μg quoted in the first printing of the DHSS (1979) tables, and more of the dements (23 per cent) then controls (3 per cent) has low serum concentrations of folate. Batata, Spray, Bolton, Higgins, and Wollner (1967) reported low serum folic acid levels and evidence of iron deficiency in patients admitted to a geriatric department. None of the patients in our study, or in that of Batata and colleagues, had megaloblastic anaemia, despite these findings. Cases of dementia due to folic acid deficiency have been reported

(Sneath, Chanarin, Hodkinson, McPherson, and Reynolds 1973).

The low plasma tryptophan concentrations, found previously by Shaw *et al.* (1981), and in this study in dementia, could have been due to stimulation of the kynurenine pathway by which tryptophan is converted to nicotinamide. However, the conversion of tryptophan to nicotinamide is minimal when the intake of nicotinic acid is adequate, as was the case for most of our subjects. Furthermore, the excretion of NMN by the female patients was lower than in the controls. If the pathway had been stimulated, the reverse would have been expected.

Thus, whilst there was evidence of dietary deficiency of a number of nutrients in our subjects, and for some nutrients more of the patients than controls had biochemical evidence of deficiency, these *on their own* (i.e. taken singly) do not correlate well with the presence of dementia. Of particular significance was the fact that the dietary intake of tryptophan did not correlate with the finding of lower plasma tryptophan concentrations in the patients. However, the finding that these dietary intakes and plasma concentrations were positively correlated in the healthy controls suggests that although the patients have a similar dietary intake, other factors, such as increased metabolic requirements or renal losses, may have been responsible for the lower plasma concentrations.

A proportion of the tryptophan in plasma is bound to albumin. Albumin concentrations in the plasma of the patients with senile dementia were significantly lower than in controls, and thus the lower concentrations of protein-bound tryptophan in the plasma of the patients could simply have been due to the lower levels of albumin available for binding. However, this could not account entirely for the finding, since statistically there was no correlation between albumin and tryptophan concentrations. The binding of tryptophan to albumin is weakened by NEFA (Curzon *et al.* 1973). and a significant correlation was noted between NEFA and the free portion of tryptophan in the plasma of the control group. The absence of such a correlation in the patients could have been due to the lower concentrations of NEFA in these individuals, and the fact that insufficient was available to effect the binding.

A number of drugs which alter NEFA plasma concentrations also alter plasma free tryptophan in the same direction (Curzon and Knott 1974). Curzon *et al.* (1973), in studies with depressed patients, found the binding of tryptophan to albumin to be affected by certain drugs. It is interesting to note that the mean NEFA concentration of those patients taking drugs in this study was lower than in those that were not on any drugs. This would have the effect of removing the interference of NEFA on binding of tryptophan to albumin. The observed lowering of plasma free tryptophan and elevation of bound tryptophan in the patients on drugs is consistent with this view. The trends are not marked, probably because, as mentioned earlier, the NEFA was not present in high enough concentrations in the plasma of patients with senile dementia to affect binding of albumin to tryptophan to any great extent.

Insulin concentrations in the plasma of patients with senile dementia were similar to those in the plasma of controls. Thus, there was no evidence to support an effect of insulin on the ratio of the concentration of tryptophan to the sum of the concentration of the five neutral amino acids which compete with it for entry into the brain (Fernstrom *et al.* 1974).

CONCLUSIONS

The starting point of this combined dietary and biochemical investigation was the finding of low plasma tryptophan concentrations in patients with senile dementia (Shaw *et al.* 1981). The results have shown that there are certain similarities and differences between these patients and age-matched controls. Thus, with reference to the RDA, the intake of both groups, particularly of some of the vitamins (thiamine, folic acid, vitamin C and vitamin D) give some cause for concern. In those instances in which intakes and nutrition of patients with senile dementia were lower than those of the controls, it is tempting to suggest that there was some inadequacy in the menu and provision of food, since the patients tended to consume all that was put before them. This, however, may not be the whole story (*vide infra*).

The interrelationship between plasma concentrations of tryptophan and other relevant metabolic measurements did not seem to provide a satisfactory explanation for low tryptophan concentrations in the patients. It remains a possibility therefore, that the explanation lies in the metabolic needs, renal losses or perhaps some degree of malabsorption as suggested by Lehmann (1981). If there were an absorptive abnormality, this may not be independent of the nutrition of the patients because, in susceptible individuals, a vicious circle could be set up in which one exacerbates the other (i.e. a nutritional deficiency undermines absorption abilities, and so on).

The general tendency for there to be a larger proportion of patients with low levels of some vitamins would be of importance if these patients had some special constellation of deficiencies, or some special vulnerability. The particular abnormalities present in patients with senile dementia were in individuals low in folate, or in concentrations of tryptophan, or in nicotinamide. Any of these three particular abnormalities may be relevant to the aetiology of dementia.

These interrelationships are worthy of further study in view of the possibility that nutritional intervention may decrease the likelihood of developing the disorder, or prevent the further development of it in individuals in whom early signs have been detected. The most promising area at this stage of research are replacement of vitamins and amino acids, and basic investigations in both areas.

Acknowledgement

We thank Dr. R. J. Ancill, Miss A. Davey, and Messrs. Bencard for their support of this study, and Dr. M. B. Briscoe and Sister L. B. Tidmarsh for their help. Our thanks also go to the volunteers who acted as controls, and to Dr. Wilson who allowed us access to his wards. In addition, we are grateful to Miss D. Hayward for help in the initial stages of this work.

Dr. D. M. Shaw is a member of the External Staff of the Medical Research Council, and Mr. S. F. Tidmarsh is also funded by the M.R.C.

References

Asplund, K., Normonk, M., and Patterson, V. (1981). Nutritional assessment of psychogeriatric patients. *Age and Ageing* **10**, 87–94.

Association of Vitamin Chemists (1966). In *Methods of vitamin assay*, (3rd edn.) (ed. M. Freed), pp. 287–344. Interscience, London.

Batata, M., Spray, G. H., Bolton, F. G., Higgins, G., and Wollner, L. (1967). Blood and bone marrow changes in elderly patients with special reference to folic acid, Vitamin B_{12}, iron and ascorbic acid. *Br. Med. J.* **2**, 667–9.

Bayoumi, R. A., and Rosalki, S. B. (1976). Evaluation of methods of coenzyme activation of erythrocyte enzymes for detection of deficiency of vitamins B_1, B_2 and B_6. *Clin. Chem.* 22, 327–35.

Bonsnes, R. W., and Taussky, H. H. (1945). On the colourimetric determination of creatinine, by the Jaffe reaction. *J. Bio. Chem.* **158**, 581–91.

Burr, M. L., Elwood, P. C., Hall, D. J., Hurley, R. J. and Hughes, R. E. (1974). Plasma and leucocyte ascorbic acid levels in the elderly. *Am. J. Clin. Nutr.* **27**, 144–151.

Curzon, G., and Friedel, J., and Knott, P. J. (1973). The effect of fatty acids on the binding of tryptophan to plasma protein. *Nature* **242**, 198–200.

—— and Knott, P. J. (1974). Effects on plasma and brain tryptophan in the rat of drugs and hormones that influence the concentration of unesterified fatty acid in the plasma. *Br. J. Pharmac.* **50**, 197–204.

Denckla, W. D., and Dewey, H., K. (1967). The determination of tryptophan in plasma, liver and urine. *J. Lab. Clin, Med.* **69**, 160–9.

Denson, K. W., and Bowers, E. F. (1961). The determination of ascorbic acid in white blood cells. A comparison of white blood cell ascorbic acid and phenolic acid excretion in elderly patients. *Clin. Sci.* **21**, 157–62.

D.H.S.S. (1979). Recommended daily amounts of food energy and nutrients for groups of people in the United Kingdom. In *Reports on Health and Social Subjects*, p.15. HMSO, London.

Dickerson, J. W. T. (1978). The interrelationships of nutrition and drugs. In *Nutrition in the clinical management of disease* (eds. J. W. T. Dickerson and H. A. Lee) pp. 308–31. Edward Arnold, London.

—— and Pao, S-K. (1975). The effect of a low protein diet and exogenous insulin on brain tryptophan and its metabolites in the weanling rat. *J. Neurochem* **25**, 559–64.

Dreyfus, P. M. (1962). Clinical application of blood transketolase determinations. *New Engl. J.. Med.* **267**, 596–8.

Duncombe, W. G. (1964). Colourimetric micro-determination of non-esterified fatty acids in plasma. *Clin. Chim. Acta.* **9**, 122–5.

Exton-Smith, A. N. (1978). Nutrition in the elderly. In *Nutrition in clinical management of disease* (eds. J. W. T. Dickerson and H. A. Lee) pp. 73–104. Edward Arnold, London.

Fernstrom, J. D., Madras, B. K., Munro, H. N., and Wurtman, R. J. (1974). Nutrtional control of the synthesis of 5-hyroxytryptamine in the brain. In *Aromatic amino acids in the brain.* (eds. G. E. W. Wolstenholme and D. W. Fitzsimons) pp. 153–66. Elsevier, Excerpta Medica - North Holland, Amsterdam.

Goldsmith, G. A., and Miller, O. N. (1967). Determination of methyl derivatives of niacin in urine. In *The vitamins* (eds P. György and W. N. Pearson) Vol. 7, pp. 137–167. Academic Press, London.

Hodkinson, H. M. (1973). The mental impairment in the elderly. *J. Roy. Coll. Physicians* (London) **7**, 305–11.

—— Rand, P., Station, B. R., and Morgan, C. (1973) Sunlight, vitamin D and osteomalacia in the elderly. *Lancet* **i**, 910–12.

Katakity, M., Webb, J. F., and Dickerson, J. W. T. (1983). Some effects of a food supplement in elderly hospitalised patients. *Human Nutr.* **37A**, 85–93.

Kay, D. W. K., Beamish, P., and Roth, M. (1964). Old age mental disorders in Newcastle upon Tyne. *Br. J. Psychiat.* **110**, 146–159.

—— Bergman, K., and Foster, E. M. (1970). Mental illness and hospital usage in the elderly. A random sample followed up *Comp. Psychiat.* **11**, 26–35.

Laurell, C–B, (1966). Quantitative estimation of proteins by electrophoresis in agarose gel containing antibodies. *Anal. Biochem.* **15**, 45–52.

Lehman, J. (1979). How to investigate tryptophan malabsorption and the value of repeated tryptophan loads. In *Origin, prevention and treatment of affective disorders* (eds. M. Schou and E. Stromgren) pp. 125–38. Academic Press, London.

—— (1981). Tryptophan malabsorption in dementia. *Acta Psychiat. Scand.* **64**, 2, 123–31.

Loh, H. S. and Wilson, C. W. M. (1971). Relationsip of human ascorbic acid metabolism to ovulation. *Lancet* i, 110–12.

Morgan, D. B. and Hullin, R. P. (1982). The body composition of the elderly mentally ill. *Hum. Nutr. Clin. Nutr.* **36C**, 439–49.

Older, M. W. J. and Dickerson, J. W. T. (1982). Thiamin and the elderly orthopaedic patient. *Age and Ageing* **11**, 101–7.

Paul. A. A. and Southgate, D. A. T. (1979). In *McCance and Widdowson's Composition of Foods.* HMSO, London.

Riley, G. J. and Shaw, D. M. (1981). Plasma tryptophan binding to albumin in unipolar depressives. *Acta Psychiat. Scand.* **63**, 165–72.

Schorah, C. J., Newill, A., Scott, D. L., and Morgan, D. B. (1979). Clinical effects of vitamin C in elderly in-patients with low blood vitamin C levels. *Lancet* i, 403–5.

Schorah, C. J. (1982). Vitamin C status in population groups. In *Vitamin C* (eds. J. H. N. Counsell and D. H. Hornig) pp. 23–47. Applied Science Publishers, London.

Shaw, D. M., Tidmarsh, S. F., Sweeney, A. E., Williams, S., Karajgi, B. M., Elameer, M., and Twining, C. (1981). Pilot study of amino acids in senile dementia. *Br. J. Psychiat.* **139**, 580–82.

Sneath, P., Chanarin, I., Hodkinson, H. M., McPherson, C. K., and Reynolds, E. H. (1973). Folate status in the geriatric population and its relation to dementia. *Age and Ageing* **2**, 177–82.

Stamp, T. C. B., Round, J. M., Rowe, D. J. S., and Haddad, T. G. (1972). Plasma levels and therapeutic efficacy of 25-hydroxycholecalciferol in epileptic patients taking anticonvulsant drugs. *Br. Med. J.* **4**, 9–12.

Vir, D. and Love, A. M. G. (1979). The nutritional status of institutionalised and non-institutionalised aged in Belfast. *Am. J. Clin. Nutr.* **32**, 1934–47.

10

Diet and migraine

M. SANDLER, JULIA T. LITTLEWOOD, AND VIVETTE GLOVER

About a quarter of all patients with migraine believe that certain dietary factors are sometimes able to initiate their attacks. Several different surveys report findings in general agreement with each other (Hanington and Harper 1968; Moffett, Swash, and Scott 1972; Dalton 1975), and we too have found that alcoholic drinks, particularly red wine, chocolate, cheese, and citrus fruits are cited by patients as the most common triggers (Glover, Littlewood, Sandler, Peatfield, Petty, and Rose 1984) (Table 10.1). Responses to these provoking substances were very strongly related, i.e. a patient who believed that he was sensitive to one of these factors usually considered himself to be sensitive to all or at least some of the others. Despite the subjective nature of this phenomenon, Olesen, Tfelt-Hansen, Henriksen, and Larsen (1981) have reported that they can reliably induce migraine attacks with red wine, in particularly susceptible individuals. Obviously, more information is necessary but, even so, a *prima facie* case appears to have been established for food stuffs with headache-initiating ability to possess some chemical property in common (Glover *et al.* 1984).

It has been suggested that patients susceptible to such provocative agents might suffer from a food allergy. In a recent study (Egger, Carter, Wilson, Turner, and Soothill 1983), the authors showed that in a group of children with migraine and other associated symptoms such as abdominal pain, a substantial proportion improved on a restricted diet and relapsed on the reintroduction of particular foods, particularly cow's milk, egg, chocolate, orange, and wheat. They suggest

TABLE 10.1 *Dietary triggering agents for migraine as identified by 1310 unselected migraine patients*

	Percentage of total sample
Alcoholic drinks	24
(red wine)	(9)
Chocolate	19
Cheese	16
Citrus fruit	9
Coffee	7
Pork	3
Dairy products	3
Eggs	2

that intolerance to such a wide range of foods points to allergic disease rather than a metabolic defect; but there seems to be no reason why many foods should not share particular chemical substances to which some subjects are sensitive. A rise in IgE or IgE antibodies was not an outstanding feature in these patients, nor was there a strong association with skinprick tests. In a different study (Merrett, Peatfield, Rose, and Merrett 1983) of 74 adult dietary migraine patients, 45 non-dietary migraine patients and 60 controls, mean IgE and IgE titres to chocolate, cheese and milk were found to be very similar in all three groups. There was no evidence of abnormal allergic response in the migraine group as a whole or in the dietary migraine patients in particular. Thus, the case that supersensitivity of certain migraine patients to particular foods is mediated by the immune system remains to be established.

It is possible, however, that the foods in question contain some chemical, or class of chemicals, to which particular subjects have an abnormal response. It is well established that certain enzyme polymorphisms can render individuals sensitive to particular foods, as with glucose-6-phosphate dehydrogenase deficiency and favism (Beutler 1983). With dietary migraine, it has come to be accepted over the last couple of decades that tyramine is the common agent and Hanington (1967) and Hanington and Harper (1968) have suggested that a deficiency in monoamine oxidase activity is the predisposing factor. They drew an analogy between dietary migraine and the 'cheese effect', the hypertensive crisis with accompanying headache that can result when patients receiving monoamine oxidase inhibiting drugs eat tyramine-containing foods, particularly cheese. However, on closer examination the analogy is not convincing (Glover *et al.* 1984). The quality of the headache is different. The order and list of triggering foods is different (Table 10.2); chocolate and citrus fruit which appear prominently in the list of migraine provocative agents, are not particularly rich in tyramine or other monoamines, An early study (Sandler, Youdim, and Hanington 1974), which found high levels of phenylethylamine in chocolate, has not been confirmed (Schweitzer, Friedhoff, and Schwartz 1975; Hurst and Toomey 1981). Even though migraine patients as a whole do possess low mean platelet monoamine oxidase activity (Sandler, Youdim, Southgate and Hanington 1970; Sicuteri, Buffoni, Anselmi and del Bianco 1972; Bussone, Giovannini,

TABLE 10.2 *Dietary factors listed in order of importance as migraine triggers or cause of side-effects with MAOI*

Migraine triggers	Cause of side-effects with MAOI
Alcohol	Cheese
(especially red wine)	Red wine
Chocolate	Marmite
Cheese	
Citrus fruit	Pickled herring
Coffee	
Pork	

Boiardi and Boeri 1977; Glover, Peatfield, Zammit-Pace, Littlewood, Gawel, Rose, and Sandler 1981) it has not been possible to find a difference between dietary and non-dietary patients (Glover, Sandler, Grant, Rose, Orton, Wilkinson, and Stevens 1977; Glover *et al.* 1981). Some authors claim that oral tyramine can trigger an attack (Hanington 1967; see Glover *et al.* 1984) while others fail to confirm it (Moffett *et al.* 1972; Ryan, 1974; Shaw, Johnson, and Keogh 1978). Migraine sufferers with low platelet monoamine oxidase activity are more sensitive to the pressor effects of intravenously administered tyramine but patients with dietary migraine as such were not different in this respect from other migraine paitents or controls (Peatfield, Littlewood, Glover, Sandler, and Rose 1983). It remains possible that, in certain patients, tyramine can contribute to a cheese sensitivity pattern, but it is unlikely to be important in sensitivity to red wine, chocolate and citrus fruit.

A different class of chemical compounds which may be involved is the phenols. In parallel with our screening of dietary migraine patients for their platelet monoamine oxidase activity levels, we studied their phenolsulphotransferase (PST) activity also (Littlewood, Glover, Sandler, Petty, Peatfield, and Rose 1982). The enzyme inactivates both endogenous and exogenous phenols by catalysing the addition of sulphate to the phenolic group (Sandler and Usdin 1981). It exists in two forms, PST M which acts on monoamine phenols such as tyramine and dopamine, and PST P which degrades phenol itself and *p*-cresol (Rein, Glover, and Sandler 1981, 1982). In a study of 17 controls, 18 non-dietary and 16 dietary migraine patients, we found that mean PST M activity was slightly reduced in the dietary migraine group, but that mean PST P activity was about half of that of controls ($p < 0.01$) (Fig. 10.1). This preliminary study both points to the possibility of dietary migraine patients having a relative inability to metabolize phenols, and suggests that phenols might conceivably be migraine triggers. The gut wall possesses the highest activity of PST in the whole

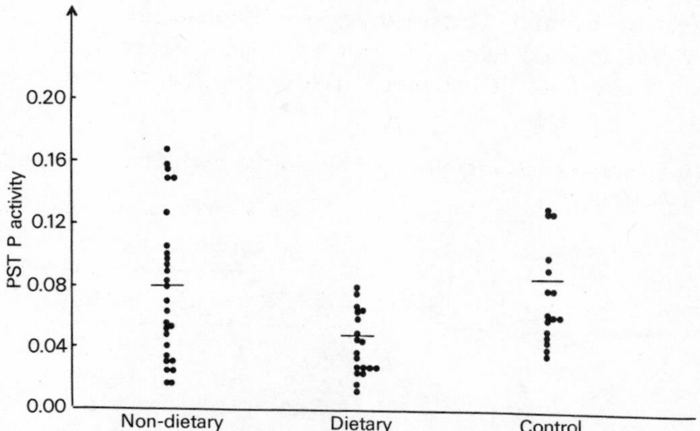

FIG. 10.1 PST P activity values in migrainous patients and controls. Activity expressed as nmoles of phenol conjugated/mg protein/10 min. $p < 0.02$: dietary less than controls; $p < 0.01$: dietary less than non-dietary; $p < 0.01$: dietary less than non-dietary plus controls. (Reproduced by kind permission of the *Lancet*, 1982, i, 984)

body (Rein *et al.* 1982), and may act as a barrier to prevent the access of noxious phenols to the circulation. If platelet activity mirrors that in the gut, then a deficit may be a marker of an attentuated gut enzyme barrier, allowing abnormal concentrations of the phenolic triggering substances to enter the bloodstream. This effect would be intensified by the action of any PST inhibitors that happened to be present in the diet.

With this in mind, we began to look for phenols in alcoholic drinks and were able to show that red wine particularly, and other alcoholic drinks to a lesser extent, contain very potent inhibitors of PST, particularly of PST P (Table 10.3) (J. T. Littlewood, V. Glover, and M. Sandler, submitted for publication). These inhibitors are sufficiently potent for half a litre of red wine to inhibit all the PST P in the body totally, whilst even one-tenth of this volume could inactivate all the PST P in the gut. Total gut PST M would be inhibited similarly by half a litre of red wine. It seems likely that these inhibitors are themselves phenols. Red wine differs from white wine primarily by being fermented in the presence of the skin and pips of the grapes and this results in the leaching out from them of both simple and complex phenols, such as catechin and the anthocyanins (Singleton and Noble 1976). Whereas red wine may contain 1200 mg/l of these compounds, white wine, typically, had only 50 mg/l in solution (Singleton and Noble 1976). Whether any of these compounds by themselves can act as migraine triggers, or whether they inhibit PST in susceptible individuals thus allowing the accumulation of other more noxious phenols, is not known,

The presence of these unknown phenolic inhibitors in food stuffs may have yet more substantial clinical implications. Normally, a large number of phenols are metabolized by sulphation, particularly at low concentration (Mülder 1982) or in individuals with low glucuronidating ability (Caldwell, Davies, and Smith 1980). Phenol itself is able to act as a cocarcinogen. Other related phenols, of dietary origin, may well possess this ability (Bakke and Midtvedt 1970) and any limiting of such inactivation might lead to increased circulating concentrations which could be harmful. Certain drugs, e.g. isoprenaline, when administered orally are metabolized almost wholly by sulphate conjugation (Morgan, Ruthven, and Sandler 1969; Conolly, Davies, Dollery, Morgan, Paterson, and

TABLE 10.3 *Percentage inhibition of phenolsulphotransferase with extracts of various alcoholic drinks*

		(mean ± SE)	
	n	PST P	PST M
Red wine	6	99.0 ± 1.3	12.0 ± 4.2
White wine	4	72.0 ± 6.5	4.8 ± 2.2
Sherry	2	80	9
Brandy	2	66	16
Whisky	3	56	5.3
Vodka	1	34	4
Gin	1	20	3

Sandler 1972). Thus, certain unexplained deaths following isoprenaline inhalation (Speizer, Doll, and Heaf 1968) may well have stemmed from failure of the sulphation system, perhaps associated with or exacerbated by concomitant ingestion of red wine. It is perfectly possible, also, that the toxic threshold for other common drugs, such as paracetamol, could be exceeded because of inhibition of its major inactivating system. It is worth noting here that certain subjects in the population who are genetically poor metabolizers (Idle, Oates, Shah, and Smith 1983) are particularly at risk from adverse drug reactions. The group of patients we have identified with dietary migraine and a deficit of PST P may well fall into this category.

References

Bakke, O. M. and Midtvedt, T. (1970). Influence of germ-free status on the excretion of simple phenols of possible significance in tumour promotion. *Experientia* **26**, 519.

Beutler, E. (1983). Glucose-6-phosphate dehydrogenase deficiency. In *The metabolic basis of inherited disease*, 5th edn (eds. J. B. Stanbury, J. B. Wyngaarden, D. S. Fredrickson, J. L. Goldstein, and M. S. Brown) pp. 1629–53. McGraw-Hill, New York.

Bussone, G., Giovannini, F., Boiardi, A., and Boeri, R. (1977). A study of the activity of platelet monoamine oxidase in patients with migraine headache or with 'cluster headaches'. *Eur. Neurol.* **15**, 157–62.

Caldwell, J., Davies, S., and Smith, R. L. (1980). Inter-individual differences in the conjugation of paracetamol with glucuronic acid and sulphate. *Br. J. Pharmac.* **70**, 112P–4P.

Conolly, M. E., Davies, D. S., Dollery, C. T., Morgan, C. D., Paterson, J. W., and Sandler, M. (1972). Metabolism of isoprenaline in dog and man. *Br. J. Pharmac.* **46**, 458–72.

Dalton, K. (1975). Food intake prior to a migraine attack – study of 2313 spontaneous attacks. *Headache* **15**, 188–93.

Egger, J., Carter, C. M., Wilson, J., Turner, M. W., and Soothill, J. F. (1983). Is migraine food allergy? *Lancet* **ii**, 865–8.

Glover, V., Littlewood, J., Sandler, M., Peatfield, R., Petty, R., and Rose, F. C. (1984). Dietary migraine: looking beyond tyramine. In *Progress in migraine research* (ed. F. Clifford Rose) Vol. 2, pp. 113–19. Pitmans, London.

—— Peatfield, R., Zammit-Pace, R., Littlewood, J., Gawel, M., Rose, F. C., and Sandler, M. (1981). Platelet monoamine oxidase activity and headache. *J. Neurol. Neurosurg. Psychiat.* **44**, 786–90.

—— Sandler, M., Grant, E., Rose, F. C., Orton, D., Wilkinson, M., and Stevens, D. (1977). Transitory decrease in platelet monoamine oxidase activity during migraine attacks. *Lancet* **i**, 391–3.

Hanington, E. (1967). Preliminary report on tyramine headache. *Br. Med. J.* **2**, 550–1.

—— and Harper, A. M. (1968). The role of tyramine in the aetiology of migraine, and related studies on the cerebral and extracerebral circulations. *Headache* **8**, 84–97.

Hurst, W. J., and Toomey, P. B. (1981). High-performance liquid chromatographic determination of four biogenic amines in chocolate. *Analyst* **106**, 394–402.

Idle, J. R., Oates, N. S., Shah, R. R., and Smith, R. L. (1983). Protecting poor metabolisers, a group at high risk of adverse drug reactions. *Lancet* **i**, 1388.

Littlewood, J., Glover, V., Sandler, M., Petty, R., Peatfield, R., and Rose, F. C. (1982). Platelet phenolsulphotransferase deficiency in dietary migraine. *Lancet* i, 983–6.

Merrett, J., Peatfield, R. C., Rose, F. C., and Merrett, T. G. (1983). Food related antibodies in headache patients. *J. Neurol. Neurosurg. Psychiat.* **46**, 738–42.

Moffett, A., Swash, M. and Scott, D. F. (1972). Effect of tyramine in migraine: a double-blind study. *J. Neurol. Neurosurg. Psychiat.* **35**, 496–9.

Morgan, C. D., Ruthven, C. R. J., and Sandler, M. (1969). The quantitative assessment of isoprenaline metabolism in man. *Clin. Chim. Acta.* **26**, 381–6.

Mülder, G. J. (1982). Conjugation of phenols. In *Metabolic basis of detoxication* (eds. W. B. Jakoby, J. R. Bend, and J. Caldwell) pp. 247–69. Academic Press, New York.

Olesen, J., Tfelt-Hansen, P., Henriksen, L., and Larsen, B. (1981). The common migraine attack may not be initiated by cerebral ischaemia. *Lancet* **ii**, 438–40.

Peatfield, R., Littlewood, J. T. Glover, V., Sandler, M., and Rose, F. C. (1983). Pressor sensitivity to tyramine in patients with headache: relationship to platelet monoamine oxidase and to dietary provocation. *J. Neurol. Neurosurg. Psychiat.* **46**, 827–31.

Rein, G., Glover, V., and Sandler, M. (1981). Phenolsulphotransferase in human tissue: evidence for multiple forms. In *Phenolsulfotransferase in mental health research* (eds M. Sandler and E. Usdin) pp. 98–126. Macmillan, London.

—— Glover, V., and Sandler, M. (1982). Multiple forms of phenolsulphotransferase in human tissues: selective inhibition by dichloronitrophenol. *Biochem. Pharmac.* **31**, 1893–7.

Ryan, Jr., R. E. (1974). A clinical study of tyramine as an etiological factor in migraine. *Headache* **14**, 43–8.

Sandler, M. and Usdin, E. (eds.) (1981). *Phenolsulfotransferase in mental health research.* Macmillan, London.

—— Youdim, M. B. H., Southgate, J., and Hanington, E. (1970). The role of tyramine in migraine: some possible biochemical mechanisms. In *Background to migraine* (ed. A. L. Cochrane) pp. 104–15. Heinemann, London.

—— Youdim, M. B. H., and Hanington, E., (1974). A phenylethylamine oxidising defect in migraine. *Nature, Lond.* **250**, 335–7.

Schweitzer, J. W., Friedhoff, A. J., and Schwartz, R. (1975). Chocolate, β-phenylethylamine and migraine re-examined. *Nature, Lond.* **257**, 256.

Shaw, S. W. J., Johnson, R. H., and Keogh, H. J. (1978). Oral tyramine in dietary migraine sufferers. In *Current concepts in migraine research* (ed. R. Greene) pp. 31–8. Raven Press, New York.

Sicuteri, F., Buffoni, F., Anselmi, B., and del Bianco, P. L. (1972). An enzyme (MAO) defect on the platelets in migraine. *Res. Clin. Stud. Headache* **3**, 245–51.

Singleton, V. L. and Noble, A. C. (1976). Wine flavour and phenolic substances. In *Phenolic, sulfur and nitrogen compounds in food flavors* (eds. G. Charalambous and I. Katz) pp. 47–70. American Chemical Society Symp. Series No. 26.

Speizer, F. E., Doll, R., and Heaf, P. (1968). Investigation into use of drugs preceding death from asthama. *Br. Med. J.* **1**, 339–43.

11

Adverse reactions to food

M. H. LESSOF

INTRODUCTION

Not all patients who have symptoms provoked by food are aware of the connection between the two. Conversely, not all patients who claim to suffer food-induced symptoms can be shown objectively to have any such reaction. If there is doubt about the diagnosis of food intolerance it may therefore be necessary to impose some form of dietary restriction which eliminates all suspect foods, and then to arrange a food challenge repeated as often as necessary, under conditions in which the response can be assessed objectively, i.e. in circumstances in which the food or a 'control' substances (given as a flavoured purée or through a naso-gastric tube) cannot be recognized by the patient. Using this approach for the more difficult cases, it is possible to distinguish between food intolerance on the one hand, and psychiatric illness on the other, occurring in individuals whose symptoms require treatment but who do not have a food-related disease.

FOOD AVERSION

The majority of adults who claim that specific foods provoke symptoms prove to have psychological problems (Table 11.1). An obsessional approach to diet characterizes some cases, and parental attitudes may on occasion affect children to the point where they develop 'intolerance by proxy'. In others, the primary disorder may concern the patient's conception of an ideal bodily configuration or weight, to achieve which a young woman with bulimia or anorexia nervosa may resolutely adjust her eating habits.

TABLE 11.1 *Food aversion*

1. Psychological intolerance:
   - neurotic fads
   - intolerance by proxy
   - 'total allergy'
2. Eating disorders:
   - anorexia nervosa
   - bulimic syndrome
   - mood disturbances

The importance of recognizing food aversion lies in the danger that misdiagnosis may lead to complex and harmful forms of treatment often involving a totally inadequate or unbalanced diet. The two most common misconceptions relate to the diagnosis of hyperactivity in childhood, and to the often unrecognized hyperventilation syndrome in adult life. While food-allergic children are occasionally emotionally disturbed or hyperactive, there is no convincing evidence that childhood hyperactivity, unaccompanied by other features of allergy or food intolerance, has anything to do with food (Joint Committee 1984). As for hyperventilation, one of the most under-diagnosed conditions in medicine, this may involve tingling of the extremities, sweating, giddiness or even loss of consciousness, provoked by a hyperventilating response to stress and easily reproduced by asking a patient to overbreathe (Nixon 1982).

FOOD INTOLERANCE

Once a diagnosis of food intolerance has been made and confirmed by re-challenge, there still remain a number of clinical syndromes involving either 'immediate' reactions within an hour, often in response to very small quantities of the offending food, or reactions of slower onset including, at the other end of the scale, delayed and insidious clinical problems such as eczema, which may occur only after a number of days or even weeks and after repeated ingestion of large quantitites of the offending food (Lessof 1983). The immediate symptoms range from gastrointestinal symptoms, including swelling of the lip or the throat, vomiting, and abdominal colic, to such other manifestations as rhinorrhoea, asthma, and urticaria. Although the time of onset may vary from case to case, a wide range of symptoms can develop more insidiously, including headache, irritability, abdominal bloating, diarrhoea, eczema, and joint pains. The different mechanisms involved are summarized in Table 11.2. Diagnostic tests remain inadequate in this area, and it is clear that detection of the irritant effects of spices or of the toxic effect of contaminated food depends entirely on the history. So does an identification of the pharmacological effects of coffee. It is possible that other pharmacological effects will be shown to correlate with detectable biochemical abnormalities but the finding of high prostaglandin $F_{2\alpha}$ levels in patients who develop urticaria, either after aspirin or after a wide range of foods, suggests some form of idiosyncrasy occurring in susceptible individuals rather than a pharmacological effect which might be obtained in the population at large (Asad, Kemeny, Youlten, Frankland, and Lessof 1984).

TABLE 11.2 *Food intolerance*

1. Irritant and toxic
2. Pharmacological
3. Enzyme defects
4. Immunological
5. Fermentation of food residues
6. Unclassified

ENZYME DEFECTS

Tests are now available for lactase deficiency, which is responsible for a substantial number of cases of milk intolerance, and for aldehyde dehydrogenase deficiency, which is responsible for alcohol intolerance in up to 40 per cent of oriental communities, and in a lower proportion of populations elsewhere. There may well be other examples of enzyme deficiencies which present as food intolerance, not only in the organic acidaemias of childhood, but also in occasional instances where, for example, severe gastrointestinal reactions to mushrooms may occur because of a deficiency of the enzyme trehalase which leads to an intolerance to the mushroom sugar, trehalose (Joint Committee 1984).

FOOD ALLERGY AND OTHER IMMUNOLOGICAL REACTIONS

Those who develop an immediate lip swelling or asthmatic reaction on ingesting small quantities of foods which are known to be rich in allergenic proteins - for example fish, egg, nuts or milk - often have supporting evidence of an IgE-mediated reaction such as a positive skin prick test or serum radioallergosorbent test against the appropriate food extract. In such cases the diagnosis is in little doubt, but where the reactions are confined to the gastrointestinal tract (Lessof, Wraith, Merrett, Merrett, and Buisseret 1980) and IgE antibody tests are negative, diagnosis may be much more difficult. In any case, the effects upon other organs of the body may depend either on gastrointestinal absorption of food antigens, which are then circulated through the body and provoke reactions elsewhere - or they may depend on the release of mediators in the bowel wall which are similarly transported to distant sites, where they affect organs which are not necessarily sensitized themselves. A number of studies have now been carried out in which the release of mediators such as prostaglandins have been monitored either in the venous blood or in rectal fliud (Jones, McLaughan, Shorthouse, Workman, and Hunter 1982). Not only is there evidence that reactions of food intolerance are accompanied by the release of prostaglandins in many cases, but it has also been shown that the use of cyclooxygenase inhibitors, such as aspirin, can in some cases prevent both prostaglandin release and the clinical manifestations of a reaction (Buisseret, Youlten, Heinzelman, and Lessof 1978). Prostaglandin rises are not always detected and, since many other mediators are also involved, it is clear that the patterns of response are more complex than was at first appreciated. Few studies involving mediator release have been published, but prostaglandins are not the only substances which deserve study. Patients with food-induced joint pains have been reported as having a significant 5-hydroxytryptamine release from platelets (Little, Stewart, and Fennessy 1983). In asthma, an indirect way of studying subclinical changes has been suggested by the observation that a changing sensitivity to bronchial challenge with histamine may develop in sensitive subjects who are given cola drinks (Wilson, Vickers, Taylor, and Silverman 1982). Whatever the explanation of these observations, there is still much to be learned about the role of mediators in food-allergic or other food-intolerant reactions. Further observations of this kind could therefore be of considerable value, both in diagnosis and in the development of pharmacological methods for counteracting reactions of this kind.

FERMENTATION OF FOOD RESIDUES

In the case of lactose-deficient patients, it has been shown that lactose residues and their fermentation in the lower bowel can be responsible for the symptoms experienced by the patient. Jones and her colleagues (1982) have suggested that the irritable bowel syndrome is not infrequently associated with intolerance to particular foods, which may have to be eaten in substantial quantities in order to produce their gastrointestinal effect. It remains to be seen whether this is another situation in which the fermentation of food residues is the cause of problems – if so, with the identification of a further mechanism for the production of food intolerant symptoms, a new approach to therapy will be required, possibly based on attempts to modify the bacterial flora.

OTHER MECHANISMS

There remain a number of instances where the mechanism of food intolerance fits poorly into any current classification, as in some patients with x-ray negative dyspepsia, whose symptoms are provoked by specific foods, and in others who have fat intolerance associated with gall stones or steatorrhoea. As in other types of food intolerance, until the mechanisms involved are better understood, appropriate food avoidance may provide the only practical approach to treatment.

PROBLEMS FOR THE FUTURE

The inconvenience and the dangers of unduly restricted diets both provide arguments for further research in this area. Better diagnostic methods may help to prevent inappropriate dietary restriction in those with psychiatric illness, so that treatment can be based upon an accurate assessment of the problem. In those who do have specific food intolerance, it is important to achieve a more scientific analysis of the mechanism involved, and of the mediators which are released. This in turn may lead to better approaches to therapy, whether pharmacological, immunological, or of some other type.

References

Asad, S. I., Kemeny, D. M., Youlten, L. J. F., Frankland, A. W., and Lessof, M. H. (1984). Effect of aspirin in 'aspirin sensitive' patients *Br. Med. J.* **288**, 745–8.

Buisseret, P. D., Youlten, L. J. F., Heinzelmann, D. I., and Lessof, M. H. (1978). Prostaglandin-synthesis inhibitors in prophylaxis of food intolerance. *Lancet* **i**, 906–8.

Joint Committe (1984). Food intolerance and food aversion. *J. R. Coll. Physic.* **18**, 83–123.

Jones, V. A., McLaughlan, P., Shorthouse, M., Workman, E., and Hunter, J. O. (1982). Food intolerance: a major factor in the pathogenesis of irritable bowel syndrome. *Lancet* ii, 1115–7.

Lessof, M. H. (1983). Food intolerance and allergy – a review. *Q. J. Med.* **206**, 111–9.

—— Wraith, D. G., Merrett, T. G., Merrett, J., and Buisseret, P. D. (1980). Food allergy and intolerance in 100 patients – local and systemic effects. *Q. J. Med.* **195**, 259–71.

Little, C. H., Stewart, A. G., and Fennessy, M. R. (1983). Platelet serotonin release in rheumatoid arthritis: a study in food-intolerant patients. *Lancet* ii, 297–9.

Nixon, P. G. F. (1982). 'Total allergy syndrome' or fluctuating hypocarbia. *Lancet* i, 404.

Wilson, N., Vickers, H., Taylor, G., and Silverman, M. (1982). Objective test for food sensitivity in asthmatic children: increased bronchial reactivity after cola drinks. *Br. Med. J.* **284**, 1226–8.

Index

RANDALL LIBRARY-UNCW
3 0490 0033152 U